Calcium-Ion Batteries

David J. Fisher

Published by **Materials Research Forum LLC**
Millersville, PA 17551, USA

Published as part of the book series
Materials Research Foundations
Volume 175 (2025)
ISSN 2471-8890 (Print)
ISSN 2471-8904 (Online)

Print ISBN 978-1-64490-348-3
ePDF ISBN 978-1-64490-349-0

This book contains information obtained from authentic and highly regarded sources. Reasonable efforts have been made to publish reliable data and information, but the authors and publisher cannot assume responsibility for the validity of all materials or the consequences of their use. The authors and publishers have attempted to trace the copyright holders of all material reproduced in this publication and apologize to copyright holders if permission to publish in this form has not been obtained. If any copyright material has not been acknowledged, please write and let us know so we may rectify in any future reprint.

Distributed worldwide by

Materials Research Forum LLC
105 Springdale Lane
Millersville, PA 17551
USA
https://mrforum.com

Printed in the United States of America
10 9 8 7 6 5 4 3 2 1

Table of Contents

Materials Research Forum LLC
https://doi.org/10.21741/9781644903490

Introduction

In the early days of their development, two types of solid-state lithium battery had utilised thin-film technology. One of these employed a $LiMn_2O_4$ thin film as the cathode, while the other was a relatively thick MoO_{3-x} cathode film. This offered a discharge capacity of $290\mu Ah/cm^2$ but it was already realised that multivalent-ion batteries with divalent ions could be more advantageous as they could offer a superior performance to that of lithium-ion batteries. Calcium-ion batteries are a particularly promising rival to the currently ubiquitous lithium-ion batteries due to the abundance of this decidedly non-strategic element and its low redox potential (close to that of lithium). They are also, unlike lithium-based batteries, unlikely to cause tragic accidents or poison the environment when clumsily disposed of.

The so-called post-lithium batteries include sodium-ion, potassium-ion, magnesium-ion, calcium-ion, aluminium-ion and zinc-ion. Interest in them is obviously driven by the limited availability of lithium and by their higher theoretical specific energies when compared with those of state-of-the-art lithium-ion batteries. Calcium is the 5th most abundant element in the Earth's crust (table 1) and - most relevantly – is 2500 times more plentiful than lithium. It is routinely available, safe, non-toxic and cost-effective. The reduction potential of calcium is high, at about $2.87V_{SHE}$. This is close to that of lithium $(3.04V_{SHE})$ and lower than that $(2.37V_{SHE})$ of magnesium. The multivalent batteries, such as magnesium-ion, calcium-ion and zinc-ion, have attracted significant attention as next-generation electrochemical energy storage devices to complement conventional lithium-ion batteries but among these, calcium-ion batteries are the least explored due to problematic reversible calcium deposition-dissolution.

The devil is however in the detail of developing a whole new menagerie of electrodes and electrolytes with which to exploit the clear theoretical advantages of calcium-ion batteries. Their practical exploitation is currently impeded by a lack of reliable electrode materials. Calcium-ion batteries are also of interest because of their high energy-density and rapid ion-diffusion, in a liquid electrolyte, which results from the bivalence and low charge/radius ratio of the calcium ion. Aqueous calcium-ion batteries exhibit reliable long-term cycling and a good rate performance but have a low energy-density due to the narrow electrochemical stability window of aqueous electrolytes. The development of compatible electrolytes is therefore also of great interest. Although calcium-ion batteries are more sustainable than lithium-ion batteries, there still remains the problem of a lack of highly rechargeable electrodes. Divalent calcium ions and reactive calcium metal interact strongly with cathode materials and non-aqueous electrolyte solutions, leading to high charge-transfer barriers at the electrode-electrolyte interface and a consequently poor electrochemical performance. The high charge-density and large radius of the Ca^{2+} ion lead to a low calcium storage capacity and an unsatisfactory cycling performance for most electrode materials. The large and divalent nature of Ca^{2+} also leads to strong interactions with intercalation hosts, sluggish ion-diffusion and a low power output. The high charge-density of divalent Ca^{2+} establishes a strong electrostatic interaction with the host lattice, resulting in low-capacity calcium-ion storage. Electrochemical systems which are based upon calcium intercalation or de-intercalation require the development of intercalation electrode materials that exhibit reversible calcium-exchange together with a reasonable energy-density and power-density. The present work summarises the research conducted so far on identifying suitable electrode and electrolyte materials. It is perforce a little chaotic, given the many possibilities which are being explored.

Table 1. Properties of competing battery elements

Element	Abundance(%)	Standard Potential(V_{SHE})	Hydrated Ion Radius(Å)
Lithium	0.0018	-3.04	3.40
Sodium	4.15	-2.71	2.76
Potassium	2.09	-2.93	2.01
Magnesium	2.33	-2.37	3.00
Calcium	4.15	-2.87	4.12
Aluminium	8.23	-1.66	4.80

Candidate Electrode Materials

Before looking for more exotic electrode materials, it is natural to consider tried and trusted choices. Attention has therefore been directed to the investigation of carbon-based materials.

Carbon-Based

graphite

Calcium-ion batteries are attractive because Ca^{2+} has the closest reduction potential to that of lithium (-2.87V_{SHE} versus -3.04V_{SHE}). This leads to a wide voltage window for a battery. Most of these have a working voltage of less than 2V, and a cycling stability of less than 50 cycles. A high-performance calcium-ion battery having a dual-carbon configuration with mesocarbon micro-beads and expanded graphite as the anode and cathode, respectively, was created[1]. It offered a reversible discharge capacity of 66mAh/g and a working voltage of 4.6V. It also exhibited good cycling stability, with a discharge capacity of 62mAh/g after 300 cycles and a capacity-retention of 94%. When graphite was used as the cathode and tin foil was instead used as the anode[2], the battery operated according to a highly reversible electrochemical reaction which combined hexafluorophosphate intercalation and de-intercalation at the cathode with a calcium-related alloying/de-alloying reaction at the anode. The optimized battery offered a working voltage of up to 4.45V and a capacity-retention of 95% after 350 cycles. Graphite (table 2) was proposed to be a suitable Ca^{2+} intercalation anode in tetraglyme[3]. The graphite could reversibly accommodate tetraglyme-solvated Ca^{2+}. When charged, the graphite formed a ternary intercalation compound, $Ca-G_4 \cdot C_{72}$, with no calcium plating. When it was charged, graphite accommodated the solvated Ca^{2+}-ions and offered a reversible capacity of 62mAh/g. This reflects the formation of a ternary intercalation compound. The mass/volume changes which occurred during intercalation, and X-ray diffraction studies, suggested that the intercalation led to the formation of an intermediate phase, between stage-III and stage-II, having a gallery height of 11.41Å. Density functional theory calculations revealed that the most stable conformation of the intercalation compound possessed a planar structure which consisted of Ca^{2+}, surrounded by tetraglyme. This eventually formed a double stack which was aligned with the graphene layers following intercalation. In spite of the marked dimensional changes which occurred during charging and discharging, the rate performance and cyclic stability remained very good. Graphite offered a capacity, at high charge and discharge rates, of the order of 47mAh/g at 1.0A/g and 62mAh/g at 0.05A/g. There was no capacity decay during up to 2000 charge and discharge cycles. A Ca^{2+}-shuttling battery was constructed by combining a graphite anode with an organic cathode, perylene-3,4,9,10-tetracarboxylic di-anhydride, in $Ca(TFSI)_2$/tetraglyme. The rocking-chair type calcium-ion battery offered a reversible capacity greater than 80mAh/g and good stability over 100 charge/discharge cycles with a reversible potential of about 1.6V and a coulombic efficiency greater than 98%.

Table 2. Present and former materials proposed as anodes for calcium-ion batteries

Anode	Electrolyte	Reversible Capacity	Final Capacity	Cycles
graphite (KS6L)	Ca(TFSI)$_2$/tetraglyme	62mAh/g	51mAh/g	2000
calcium metal	Ca(BF$_4$)$_2$/carbonates	21mC	15mC	30
calcium metal	Ca(TFSI)$_2$/monoglyme	80mAh/g	-	few
lithium metal	LiPF$_6$/Ca(PF$_6$)$_2$/*	95mAh/g	75mAh/g	1500
gold (substrate)	Ca(BH$_4$)$_2$/THF	1mAh/cm^2	-	-
tin foil	Ca(PF$_6$)$_2$/carbonates	530mAh/g	-	-
tin foil	Ca(PF$_6$)$_2$/**	86mAh/g	70mAh/g	1000

*EC+EMC+PC+DMC, **EC/PC/DMC/EMC, EC: ethylene carbonate, EMC: ethyl methyl carbonate, PC: propylene carbonate, DMC: dimethyl carbonate, THF: tetrahydrofuran, TFSI: trifluoromethanesulfonyl)imide

A graphite electrode could offer a rate capability of up to 2A/g, with some 75% of the specific capacity at 50mA/g. Full calcium intercalation corresponded to about 97mAh/g. The capacity was retained after 200 cycles, without any appreciable cyclic degradation. The calcium ions were intercalated into the graphite via a staging process. The intercalation mechanism was clarified by means of synchrotron X-ray diffraction, nuclear magnetic resonance and first-principles calculations[4]. Electrochemical intercalation of Ca^{2+} into a graphite electrode was found[5] to be possible when γ-butyrolactone was used as a solvent, and this resulted in a reversible charge/discharge capacity. The γ-butyrolactone-based electrolyte permitted a reversible redox reaction, with intercalation and de-intercalation of Ca^{2+} within the electrode. On the other hand, Ca^{2+} could not be intercalated between graphite layers in an ethylene carbonate plus diethyl carbonate electrolyte. The interaction between the γ-butyrolactone solvent and Ca^{2+} was weak while the interaction between the ethylene carbonate plus diethyl carbonate electrolyte and Ca^{2+} was strong. Analysis of the surface of the negative electrode, following charging and discharging, showed that a component which appeared to be a solid electrolyte interphase was present in the graphite electrode when using the γ-butyrolactone-based electrolyte.

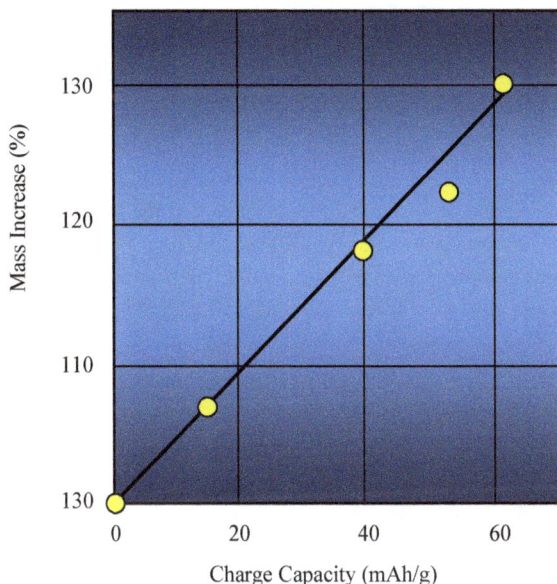

Figure 1. Mass increase of graphite as a function of charge capacity

Bacterial cellulose, with a highly graphite-like ordered structure, was proposed[6] as an efficient co-intercalation host for multivalent calcium ions. A free-standing bacterial cellulose electrode, having a 3-dimensional porous morphology, exhibited ultra-fast co-intercalation kinetics. By using bacterial cellulose electrodes having various crystallinities, it was found that the co-intercalation depended directly upon the degree of carbon crystallinity.

A 3.5M concentrated electrolyte was developed[7] which consisted of a calcium bis(fluorosulfonyl)imide salt that was dissolved in carbonate solvents. This electrolyte greatly improved the intercalation capacity of anions in the graphite cathode, and aided the reversible insertion of Ca^{2+} into the organic anode. When this concentrated electrolyte was combined with a graphite cathode and organic anode, the resultant calcium-based dual-ion battery offered a specific discharge capacity of 75.4mAh/g at 100mA/g, with 84.7% capacity-retention after 350 cycles.

The electrochemical intercalation of Ca^{2+} or solvated Ca^{2+} into Marimo nano-carbon was studied[8] with regard to its use as an anode for calcium-ion batteries. When $Ca(ClO_4)_2$ and

Ca(TFSI)$_2$ were used as electrolyte salts, the solution structure changed in γ-butyrolactone solvent. A higher charge-capacitance was obtained at the Marimo nano-carbon electrode than at the graphite electrode. When heat-treated Marimo nano-carbon was used, the (ClO$_4$)$_2$ intercalated better than did Ca(TFSI)$_2$.

Further study[9] clarified the critical role which anions play in modulating calcium-ion solvation structures, and their intercalation into graphite. Electrostatic interactions between calcium ions and anions were found to govern the configurations of solvated calcium ions in dimethylacetamide-based electrolytes and graphite intercalation compounds. Various anions were considered, and it was found that a coordination with 4 solvent molecules per calcium ion led to the highest reversible capacities and the fastest reaction kinetics. This rationalised the cause of the various calcium-ion intercalation behaviours which were observed among the various anion-modulated electrolytes.

graphene

Hydrogenation improved the performance of graphitic materials which were used as anodes in lithium- and sodium-ion rechargeable batteries, and density functional theory calculations indicated that this was also true of sodium- and calcium-ion batteries[10]. This could be attributed to an increase in the binding strength of the metal adatom to the hydrogenated graphene and to an increase in the interlayer spacing of the layered materials. It was estimated that sodium and calcium were bonded weakly to graphene sheets, with binding energies of -0.763 and -0.817eV, respectively. They were bound more strongly to a single layer of hydrogenated graphene sheet (C$_{68}$H$_4$), with binding energies of -1.670 and -2.756eV, respectively. Although the sodium was not strongly intercalated into the layers of C$_{68}$H$_4$, calcium could intercalate into this material to impart an electrical capacity of 591.2mAh/g, with a 29.3% expansion of the interlayer distance. The hydrogenated defective graphene was thus an anode material that might enhance the performance of rechargeable batteries. The effect of vacancies, with and without hydrogen, upon the binding of calcium atoms to various defective carbon-based 2-dimensional sheets was considered (table 3). If the defects resulted in under-coordinated carbon atoms, the binding of the metal atoms could become so large that desorption was difficult. Such materials were likely to result in the linking of adjacent layers, thus impairing battery performance. Material with hydrogenated mono-vacancies had apical C-H bonds on the basal plane, and this resulted in an increase in the spacing between layers together with an increased binding of calcium, as compared with that of pristine graphene. Calcium alone provided a strong enough binding to the bulk layers of 4(H1-MVG) to prevent cohesion of the metal atoms. The bulk material provided a maximum capacity of 16 calcium atoms within each AA stacked layer; equivalent to C$_{68}$H$_4$Ca$_{16}$. The expansion of the stacked layers following calcium intercalation varied from 6.6% to 31.0%. The improved performance of hydrogenated anode materials, compared to that of pristine graphene, was attributed to a stronger binding of the metal atoms to the substrate and to the expansion of the material before adding metal atoms. This expansion was due mainly due to apical C-H bonds which acted as a side-arm to expand the layers and enable intercalation.

Table 3. Lattice vector lengths of graphene sheets with various vacancies and difference with respect to graphene

Material	Vector Length(Å)	Difference(%)
graphene	14.81	-
mono-vacancy graphene	14.77	-0.3
di-vacancy graphene	14.67	-0.9
tri-vacancy graphene	14.65	-1.1
quad-vacancy graphene	14.54	-1.8
24-vacancy graphene	22.27	0.3

General design principles were proposed[11], for graphene anodes for batteries, on the basis of density functional theory calculations and experimental studies of heteroatom-doped graphene. This made it possible to choose the best dual-doped graphene anode; one which offered a 10-fold higher Ca^{2+} storage capability than that of a singly-doped graphene anode. The universal design principle went beyond the Conway pseudo-capacitive theory to describe the charge-storage mechanisms of pseudo-capacitive materials. The so-called descriptor was based upon the intrinsic physical properties of dopants and correlated doping structures with capacitive properties. It offered a deeper understanding of the pseudo-capacitive mechanism. The descriptor (figures 2 and 3) took account of 3 intrinsic physical parameters of the dopants, which was described by $(E_X R_X/A_X)/(E_C R_C/A_C)$, where A_X, R_X and E_X were the electronegativity, radius and electron affinity, respectively, of the dopant, X, and A_C, R_C and E_C were the electronegativity, radius and electron affinity, respectively, of carbon atoms.

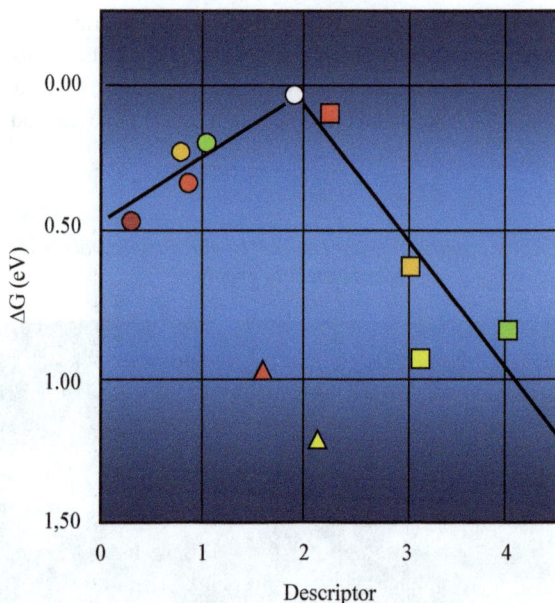

*Figure 2. Gibbs free-energy change of doped-graphene as a function of the descriptor.
Brown circle: boron, red circle: nitrogen, orange circle: oxygen, green circle:
phosphorus, white circle: antimony, red square: silicon, orange square: chlorine, yellow
square: bromine, green square: iodine, red triangle: fluorine, yellow triangle: sulphur*

This descriptor could describe well the interaction between Ca^{2+} and a heteroatom-doped graphene anode. It also quantified the influence of interfacial effects upon charge storage. Plots of the descriptor had a so-called volcano shape (figures 2 and 3). The closer that a point was to the top of the volcano plot, the better was the electrochemical performance of a doped structure. Antimony-doped graphene, located at the top of the volcano, was therefore expected to exhibit the best electrochemical performance among metal-free singly heteroatom-doped graphene. The electrochemical performance of fluorine- and sulphur-doped structures was not predicted well by the descriptor. This was attributed to the fact that fluorine had the highest electronegativity. It could form various doping structures. Although the fluorine and sulphur dopant predictions deviated from the volcano relationship, this was deemed not to impair the predictive value of the descriptor.

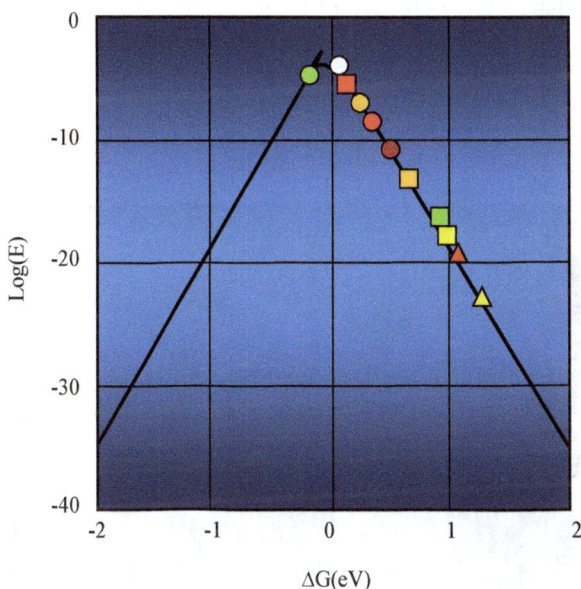

Figure 3. Energy density of doped-graphene as a function of the Gibbs free-energy change. Brown circle: boron, red circle: nitrogen, orange circle: oxygen, green circle: phosphorus, white circle: antimony, red square: silicon, orange square: chlorine, yellow square: bromine, green square: iodine, red triangle: fluorine, yellow triangle: sulphur

The potential use of hexa-peri-hexabenzocoronene nano-graphene as an anode material in magnesium- and calcium-ion batteries was investigated[12] by means of density functional theory calculations. The magnesium or calcium cations were chemically adsorbed over the centres of rim hexagons, with adsorption energies (table 4) of -190.3 or -118.2kcal/mol, respectively. The magnesium or calcium atoms were physically adsorbed on the hexa-peri-hexabenzocoronene and released 4.1 or 4.8kcal/mol, respectively. The energy barrier which had to be surmounted by the magnesium and calcium cations in order to migrate from the surface of one hexagon, to another, was 4.2 or 5.3kcal/mol. The small values increased the ion-mobility and the charge/discharge rate. The specific capacity of hexa-peri-hexabenzocoronene nano-graphene was predicted to be 593.5 and 486.8mAh/g for magnesium and calcium atoms, respectively. A stronger cation-π interaction between Mg^{2+} and the hexa-peri-hexabenzocoronene generated a higher cell voltage for magnesium-ion batteries (3.97V), as compared to that for calcium-ion batteries (2.45V).

Table 4. Adsorption energies of calcium and Ca^{2+} on pristine and doped hexabenzocoronene

Material	E_a(kcal/mol)	E_{HOMO}(eV)	E_{LUMO}(eV)	E_g(eV)	ΔE_g(%)
HBC	-	-5.23	-1.65	3.58	-
HBC:Ca	-34.9	-3.57	-2.09	1.48	-58.7
HBC:Ca^{2+}	-143.6	-11.05	-9.58	1.47	-58.9
N-HBC	-	-3.76	-1.94	1.82	-
N-HBC:Ca	-3.4	-3.66	-1.85	1.80	-0.9
N-HBC:Ca^{2+}	-199.2	-9.51	-8.56	0.95	-47.8
BN-HBC	-	-5.76	-1.84	3.92	-
BN-HBC:Ca	-9.01	-3.70	1.91	1.79	-54.2
BN-HBC:Ca^{2+}	-109.9	-11.10	-9.69	1.41	-63.9
AlN-HBC	-	-5.47	-2.20	3.27	-
AlN-HBC:Ca	-16.4	-5.16	-2.56	2.60	-20.4
AlN-HBC:Ca^{2+}	-138.9	-11.19	9.03	2.16	-34.0

ΔE_g: change in E_g of nanographene after Ca/Ca^{2+} adsorption

The possibility of using hexa-peri-hexabenzocoronene nano-graphene and its doped analogues as electrodes in calcium-ion batteries was further investigated[13] by means of density functional theory. This showed that doping with electron-rich nitrogen atoms appreciably increased the cell voltage of hexa-peri-hexabenzocoronene. On the other hand, BN-doping did not affect the cell voltage. The cell voltages which were exhibited by the nanographenes were: N-HBC (1.01V) > AlN-HBC (0.63V) > HBC (0.56V) > BN-HBC (0.52V).

A 2-dimensional so-called Dirac material, termed graphene+, was theoretically predicted[14] to offer good stability, high ductility and marked electronic conductivity. It was therefore expected to be an ideal anode material for calcium-ion batteries (table 5). The graphene+ was predicted to offer a capacity of 1487.7mAh/g, a diffusion barrier of 0.21eV and an average open-circuit voltage of 0.51V. This was expected to affect electrolyte salvation, calcium-ion adsorption and migration. When in contact with electrolyte solvents, graphene+ was expected to exhibit a high adsorption strength and a rapid migration of calcium ions on its surface. First-principles calculations showed that individual calcium ions preferred to be adsorbed at the H_1 sites on the substrate, with an adsorption energy of -1.49eV, reflecting the strong binding between calcium and graphene+. The graphene+ monolayer exhibited a lattice shrinkage of 0.6% following the full intercalation of calcium ions.

Table 5. Possible 2-dimensional anode materials for calcium-ion batteries

Material	Specific Capacity(mAh/g)	Diffusion Barrier(eV)	OCV(V)
graphene+	1487.7	0.21-0.77	0.51
Ti_3C_2	319.8	0.12	0.09
penta-graphene	1487.6	0.36-0.58	-
WS_2	326.09	0.185-0.284	1.26
TiS_2	957.2	0.42	0.66
borophene	800	0.33	-
BC_3	783	0.147-0.571	-
MoO_2	1256	0.22	0.35
FeSe	631	1.53	0.08
VO_2	260	0.306	2.82
V_2CSe_2	394.12	0.24	0.072

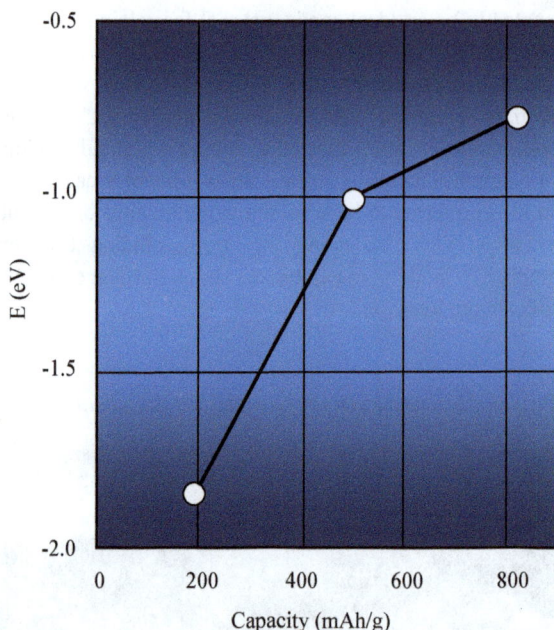

*Figure 4. Adsorption energy for pyrazinoquinoxaline graphdiyne as a function of the
storage capacity of calcium*

The adsorption and diffusion of calcium ions was studied, using density functional theory
calculations, at the van der Waals interface of 2-dimensional heterostructures which were
constructed by vertically stacking $NbSe_2$ monolayers and graphene[15]. This showed that
calcium could be effectively adsorbed at the van der Waals interface of the 2-dimensional
heterostructure, with the binding energy of the most stable site being -2.77eV. This was
much higher than that for most metal ions when bound to pristine graphene. The $NbSe_2$-
graphene 2-dimensional heterostructure therefore augmented the binding of calcium ions
at the interface. Due to the random stacking of the $NbSe_2$ and graphene, multi-path
minimum-energy paths were identified in the van de Waals region, with diffusion barriers
of 0.20 to 0.50eV. This revealed the capability of 2-dimensional heterostructures to
promote rapid multivalent ionic mobility and charge-discharge rates, while maintaining
strong binding at the van der Waals interface.

A simple process was described[16] for the construction of 2-dimensional porous hybrid nano-sheets which comprised ultra-thin (about 0.5nm) antimony nano-plates that were anchored to acetone-derived graphene-like porous carbon nano-sheets via an aldol reaction and a carbothermal reduction method which involved acetone and antimony acetate. The structural and compositional characteristics of the 2-dimensional porous hybrid nano-sheets boosted electrochemical sodium storage with regard to cycling stability, reversible specific capacity and rate performance.

graphdiyne

Two-dimensional materials such as graphdiyne have been considered as anodes, and pyrazinoquinoxaline graphdiyne nano-sheets have been proposed for use in lithium-, sodium-, calcium- and magnesium-ion batteries. Investigations were based upon first-principles electronic structure simulations. Pyrazinoquinoxaline graphdiyne offers favourable electrode properties such as storage capacities of 1938, 1716 and 830.60mAh/g for lithium, sodium and calcium ions, respectively[17]. Calculations were made of the adsorption energy (figure 4), open-circuit voltage, density of states and charge capacity.

miscellaneous carbon forms

The reversible electrochemistry of a calcium-ion cell was combined[18] with an aqueous electrolyte consisting of 1M $Ca(ClO_4)_2$. Carbon cloth, barium hexacyanoferrate (BaHCF) and meso-carbon micro-beads were essayed as the current-collector, cathode and anode, respectively. The cell provided a 40mAh/g capacity at a 5C rate up to 100 cycles.

In situ formed poly(anthraquinonyl sulphide) and carbon-nanotube composites were considered[19] as non-aqueous calcium-ion battery cathodes. The enolization redox chemistry of the organics exhibited rapid redox kinetics, and the introduction of carbon nano-tubes accelerated electron transportation and made a contribution to rapid ionic diffusion. When the conductivity of the organic was increased by an increased nano-tube content, the voltage-gap was appreciably reduced. The composite electrodes had a specific capacity of 116mAh/g at 0.05A/g, a rate capacity of 60mAh/g at 4A/g and an initial capacity of 82mAh/g at 1A/g, with 83% capacity-retention after 500 cycles. The electrochemical mechanism was such that the organic underwent reduction reaction of the carbonyl bond during discharge and became coordinated by Ca^{2+} and $Ca(TFSI)^+$ species. Computational simulations indicated that the construction of Ca^{2+} and $Ca(TFSI)^+$ co-intercalation in the organic furnished the most reasonable pathways.

Promising criteria for the choice of battery electrodes include 2-dimensionality, low density, flexibility, hydrophilicity, high metal surface diffusivity, high conductivity and mechanical strength. Attention was therefore paid[20] to biphenylene, a 2-dimensional non-benzenoid carbon allotrope that is created by bottom-up on-surface interpolymer dehydrogenation. An investigation was made of factors such as the electronic, mechanical and electrochemical properties of pristine and boron-doped biphenylene nano-sheets, such as the binding-strength, ionic diffusion barrier, equilibrium voltage and theoretical capacity. Assuming a favourable adsorption energy and no structural deformation, all of the materials with adsorbed lithium, potassium and calcium atoms exhibited a high structural stability. The ionic diffusion barrier was simulated by using a charged-electrode model which took account of possible charge-transfer polarization. It was found that the ionic diffusion barrier depended upon the surface atomic configuration. This in turn was affected by the bond length, valence-electron number, electrical conductivity, ionic diffusion barrier and equilibrium voltage. As they exhibited ion-diffusion barriers along furrows of 0.23eV, 0.21eV and 0.66eV for lithium, potassium and calcium, respectively, the structures offered good overall rate capacities. The theoretical capacity (1501.7mAh/g) was up to 4 times higher than that (372mAh/g) in the case of lithium-ion batteries with graphite, 938.5mAh/g in the case of potassium-ion batteries, and 1126.3mAh/g in the case of calcium-ion batteries.

A free-standing electrode of ZnHCF, grown on carbon nano-tube fibre, exhibited a good mechanical deformability and electrochemical behaviour for use in an aqueous fibre-shaped calcium-ion battery[21]. Due to an unique Ca^{2+}/H^+ co-insertion mechanism, this cathode could offer a good ion-storage capability within a broad voltage window. By combining it with a polyaniline on carbon-fibre anode, a battery was created which offered a volumetric energy density of 43.2mWh/cm^3 and which maintained good electrochemical properties during deformation.

An all-organic dual-ion battery, with acetonitrile plus calcium perchlorate as the electrolyte, was constructed[22]. Electrochemical energy was stored via the association and disassociation of calcium and perchlorate ions in a perylene diimide-ethylene diamine and carbon-black composite anode and a polytriphenylamine cathode with highly reversible redox states. Using this energy-storage mechanism, the electrolyte offered a good electrochemical performance at 25 to -50C. At -50C, it preserved some 61% of the capacity at 25C (83.4mAh/g) with a current density of 0.2A/g. The battery exhibited a good cycling stability at low temperature and retained 90.3% of the initial capacity at 1.0A/g after 450 charge-discharge cycles at -30C. Impedance studies which were performed at various temperatures showed that the low-temperature performance depended more upon the electrode materials than upon the electrolyte.

organic materials

A covalent organic framework was prepared[23] from 2,5-diaminohydroquinone dihydrochloride and 1,3,5-triformylphloroglucinol and was used as an anode material for calcium-ion storage in aqueous electrolytes. The framework offered a specific capacity of up to 119.5mAh/g at 1A/g and a rate performance of 78.7mAh/g, even at a current of 50A/g. Fourier-transform infra-red spectroscopy and X-ray photo-electron spectroscopy revealed the occurrence of a reversible calcium insertion and extraction behaviour, with carbonyl being the active site. A notable rate performance arose from the storage mechanism of the pseudo-capacitance behavior. The co-insertion of protons and calcium ions, and *in situ* conversion of C-OH groups, of the framework, to C=O groups was observed. A full cell was prepared with the covalent organic framework and activated carbon as the active materials. The cell offered an average working voltage of 0.6V, with 73.7% capacity-retention (0.0164% reduction/cycle over 1600 cycles).

Organic materials are promising candidates for cation storage in calcium-ion batteries, even though the high solubility of organic materials in an electrolyte and their poor electronic conductivity remain the chief barriers to the formulation of high-performance batteries. A nitrogen-rich covalent organic framework with multiple carbonyl groups was designed[24] as an aqueous anode which could overcome those problems. This organic offered a reversible capacity of 253mAh/g at 1.0A/g and a cycle life with a 0.01% capacity-decay per cycle at 5A/g after 3000 cycles. The occurrence of a redox mechanism which involved Ca^{2+}/H^+ that was co-intercalated in the organic and which was chelated with C=O and C=N active sites was confirmed. A C=C active site was identified as being involved in Ca^{2+} ion storage. Calculations and experimental data suggested that up to nine Ca^{2+} ions per organic repetitive unit were stored following 3 staggered intercalation steps that involved 3 distinct Ca^{2+} ion-storage sites.

Anodes of 3,4,9,10-perylenetetracarboxylic di-imide were used[25] in aqueous calcium-ion batteries with a water-in-salt electrolyte. The organic material offered a discharge-capacity of 131.8mAh/g, a rate-performance of 86.2mAh/g at 10000mA/g and a lifetime of 68000 cycles over 470 days. The capacity-retention was 72.7%. The calcium storage mechanism involved an enolization reaction. By matching a high-voltage cathode of Prussian Blue analogue, aqueous calcium-ion cells exhibited an operating range of -20C to 50C and a lifetime of 30000 cycles. An aqueous calcium-ion pouch cell could be constructed which offered a lifetime of more than 500 cycles.

Anodes were made of the small molecular organic material, tetracarboxylic di-imide, which underwent carbonyl enolization (CO↔CO-) in aqueous electrolytes[26]. This avoided the diffusion difficulties of intercalation-type electrodes and the limited capacity of polymer organic electrodes, leading to high and rapid calcium storage. In the case of an aqueous calcium-ion cell, the di-imide anode offered a reversible capacity of 112mAh/g, a capacity-retention of 80% following 1000 cycles and high power capability at 5A/g. It was deduced that the calcium ions diffused along the a-axis tunnel so as to enolize carbonyl groups without becoming trapped in the aromatic carbon layers. The feasibility of the present anodes was demonstrated by combining them with cost-effective Prussian Blue analogous cathodes and a $CaCl_2$ aqueous electrolyte.

A covalent organic framework having repeated pyrazine and pyridinamine units was used as an anode for calcium-ion batteries[27]. This resulted in a markedly flat ultra-low potential plateau which ranged from -0.6 to -1.05$V_{Ag/AgCl}$. This was attributed to the high level of the lowest unoccupied molecular orbital. The anode offered a rate-performance of 152.3mAh/g at 1A/g, long-term cycling stability and a capacity-retention of 89.9% following 10000 cycles. It was deduced that C=N active sites reversibly trapped Ca^{2+} ions via chemisorption during discharging and charging. The anode exhibited exceptional structural stability during cycling. By combining it with a high-voltage manganese-based Prussian Blue cathode, an aqueous calcium-ion battery with a voltage interval of 2.2V was created which offered 83.6% retention following 10000 cycles.

Density functional theory computations were made[28] of bilayer s-tetrazine-based covalent organic framework materials for use as anode materials for sodium- and calcium-ion batteries. Electronic band-structure calculations suggested that the bilayer material is an indirect band-gap semiconductor with a band-gap of 0.95eV. Sodium and calcium atoms were adsorbed on the bilayer material at the most energetically favourable adsorption sites, with adsorption energies of -1.37eV and -2.27eV, respectively. The diffusion energy barriers for the migration of sodium and calcium atoms on the bilayer material were 0.19eV and 0.63eV, respectively, thus promising rapid ion-mobility and high charge/discharge rates. The theoretical specific capacity of the associated batteries was 618.69 and 412.46mAh/g, respectively. The average voltages of the bilayer as an electrode material for sodium and calcium batteries were 0.96 and 1.13V, respectively. It was concluded that the bilayer material was suitable as an anode material for the batteries.

First-principles calculations were used[29] to evaluate anthracene, tetracene and pentacene (table 6, figure 5) as anode materials for calcium-ion batteries. The adsorption of calcium atoms on isolated molecules of any of these materials was energetically favourable, as was intercalation into bulk crystals for a wide range of calcium concentrations. For each material, the volume expansion during intercalation was less than 20%; in the case of pentacene it was less than 8%. The barriers to calcium diffusion along the polyacene molecules were less than 0.45eV, but the calcium diffusion was limited by so-called jumps between the molecules.

Table 6. Lattice parameters of polyacene crystals

Polyacene	Source	a(Å)	b(Å)	c(Å)	α(°)	β(°)	γ(°)	Volume(Å³)
anthracene	experiment	8.55	6.01	11.17	90	124.59	90	473.16
tetracene	experiment	6.05	7.83	13.01	77.12	72.11	85.79	572.97
pentacene	experiment	6.27	7.71	14.44	76.75	88.01	84.52	677.32
anthracene	calculated	8.39	5.86	11.05	90	125.24	90	444.04
tetracene	calculated	6.03	7.32	12.62	78.71	73.05	85.57	523.17
pentacene	calculated	6.27	7.55	14.18	77.65	88.83	83.93	652.17

Sequential band-filling upon increasing the level of intercalation led to re-entrant semiconducting-metallic-semiconducting behaviour. Trends in the intercalation energies were related to charge-transfer from calcium atoms to the polyacene molecules, and the intercalation-energy maximum one calcium atom per formula unit was associated with the maximum charge donation from calcium atoms, which was 1.49e and 1.46e for pentacene and tetracene, respectively. For anthracene, the charges on 1 and 2 calcium atoms in the unit-cell (0.5 and 1 atom/formula-unit) were essentially equal, at 1.45e and 1.44e, respectively. With increasing concentration, the energy-gain decreased monotonically, accompanied by a decrease in charge-transfer from calcium atoms. The calculated maximum volume expansion was 17% for tetracene and less than 10% for anthracene and pentacene. The volume changed non-monotonically. At a calcium concentration of 0.33 to 0.60, the volume increased due to the repulsion of calcium atoms and polyacene molecules. At concentrations of 0.60 to 0.66, the volume decreased slightly due to an increase in total bonding between calcium and polyacene. At high concentrations saturation occurred, charge-transfer from calcium atoms and energy-gain decreased and repulsion increased. The volume of the crystal therefore increased. At a calcium concentration greater than 0.6 in anthracene and greater than 0.66 in tetracene or pentacene, the theoretical capacity of calcium-intercalated polyacenes exceeded the theoretical capacity of lithium-intercalated graphite; a notable rival anode material for use in ion batteries. At the highest concentration at which the structures were thermodynamically stable, the theoretical capacity was about twice that of lithium-graphite. The band-gaps of anthracene, tetracene and pentacene were 1.92eV, 1.19eV and 0.85eV, respectively.

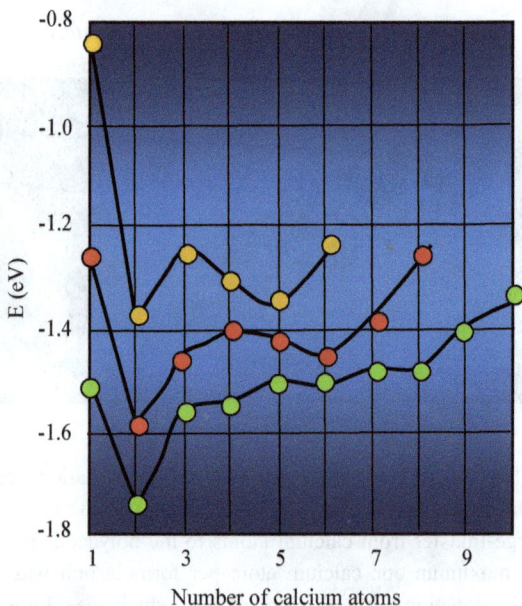

Figure 5. Energy required to take a calcium atom from the bulk crystal and adsorb it on a free-standing polyacene molecule. Anthracene: orange, tetracene: red, pentacene: green

The pristine polyacenes had 2 nearly-degenerate low-energy conduction bands which were separated by 1 to 1.5eV from the higher bands. When a calcium atom was added to the unit cell, it donated 2 electrons which then filled the lower of the 2 nearly-degenerate lowest-energy conduction bands, thus making the crystal a very narrow-gap semiconductor (0.08eV for pentacene) or a metal (anthracene, tetracene). When a second calcium atom was added, another 2 electrons were added and filled the second-lowest conduction band, producing a wider band-gap of 0.4, 0.77 and 0.86eV for anthracene, tetracene and pentacene, respectively. If yet more calcium atoms were intercalated, the system again became a narrow-gap semiconductor or metal. Diffusion was governed by so-called jump between the molecules. The diffusion barriers to the jumps were 0.68eV for anthracene, 0.92eV for tetracene and 0.8eV for pentacene. The high intercalation energies, low volume expansions and suitable kinetic and electronic properties made these polyacene molecules promising anode materials for calcium-ion batteries, with pentacene being particularly noteworthy because of its low volumetric expansion and favourable electronic properties.

The small-molecule organic material, 3,4,9,10-perylenetetracarboxylic di-imide, was investigated[30] as a calcium-ion host for calcium-ion batteries with a non-aqueous electrolyte. Structural and computational studies revealed that this material, with an increased degree of Ca^{2+} storage, could increase the solubility of reduced 3,4,9,10-perylenetetracarboxylic di-imide, due to a reduced π-π interaction which was suppressed by the saturated electrolyte. This was attributed to a high re-deposition rate to form a 3,4,9,10-perylenetetracarboxylic di-imide film with strengthened hydrogen bonds which facilitated a fast enolization reaction for Ca^{2+} storage. A cell which was assembled using a 3,4,9,10-perylenetetracarboxylic di-imide negative electrode and carbon-based positive electrode offered a power-density greater than 3000W/kg, an energy-density of about 150Wh/kg and a reversible capacity of 80mAh/g at 5A/g. The battery had a rate-performance of 90mAh/g at 1A/g and stable cyclability over 4000 cycles, with negligible decay at -10C.

An arsenic-based covalent organic framework was designed[31] by exploiting the geometrical symmetry of a semiconducting phosphazene-based covalent organic framework, with p-phenylenediamine as a linker, and hexachorocyclotriphosphazene as an arsenic-containing monomer in a C3-like spatial configuration. The arsenic-based covalent organic framework, with its engineered nanochannels, exhibited a marked anodic behavior for potassium- and calcium-ion batteries. It offered a storage capacity of about 914mAh/g, a diffusion barrier of 0.12eV, an open-circuit voltage of 0.23V and a volume expansion of 2.41% for potassium ions. The corresponding properties for calcium were 2039mAh/g, 0.26eV, 0.18V and 2.32%.

The organic crystal electrode, 5,7,12,14-pentacenetetrone, was proposed[32] as a material for aqueous calcium-ion storage. The weak π-π stacked layers of the 5,7,12,14-pentacenetetrone molecules constituted a flexible and strong structure which was suitable for calcium-ion storage. Channels within the 5,7,12,14-pentacenetetrone crystal provided good pathways for rapid ionic diffusion. A 5,7,12,14-pentacenetetrone anode offered a specific capacity of 150.5mAh/g at 5A/g, a rate capability of 86.1mAh/g at 100A/g and a favourable low-temperature behaviour. Detailed study identified proton-assisted uptake and removal of Ca^{2+} in the 5,7,12,14-pentacenetetrone during cycling. First-principles calculations suggested that the calcium ions remained within in the interstitial space of the 5,7,12,14-pentacenetetrone channels and were stabilized by carbonyl groups from adjacent 5,7,12,14-pentacenetetrone molecules. Combination of the anode with a high-voltage positive electrode led to a full aqueous calcium-ion cell.

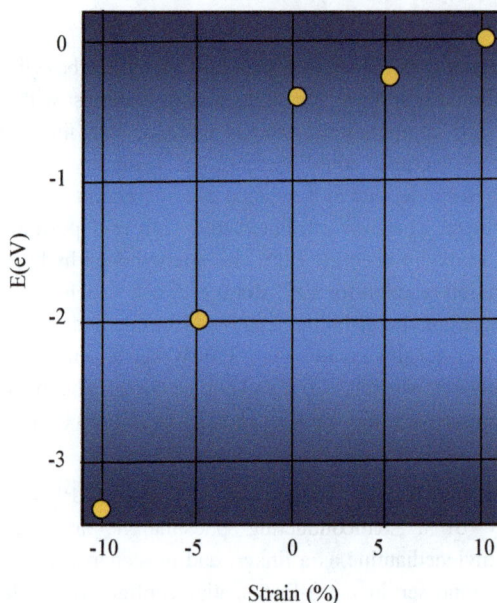

Figure 6. Calcium binding energy as a function of biaxial strain of 2-dimensional Sc$_2$C

Carbides

boron

Two-dimensional BC$_3$ monolayers have been considered[33] as candidate electrode materials due to their mechanical and chemical properties and their large surface area to volume ratio. The possibility of using the monolayers as an anode material in calcium-ion batteries was investigated by means of first-principles density functional theory calculations. These gave a calcium storage capacity of 783mAh/g and an open-circuit voltage of 0.027V. The adsorption energy of calcium was -2.59eV, and the migration energy barrier of a single calcium atom was 0.147eV. It was concluded that BC$_3$ monolayers could be used as 2-dimensional anode materials.

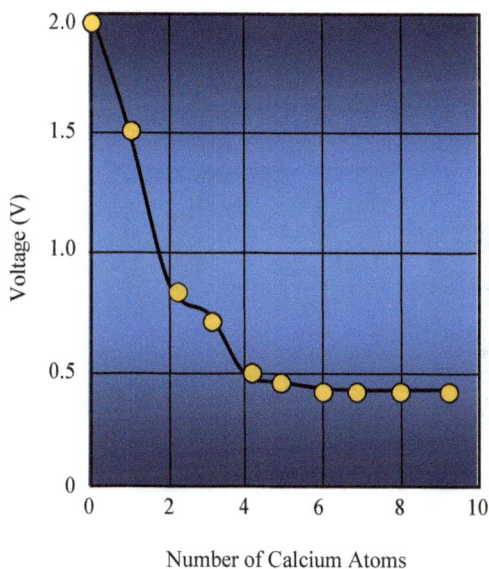

Figure 7. Open-circuit voltage of SiC as a function of number of calcium atoms in the initial adsorption layer

scandium

Density functional theory calculations were made[34] of calcium-binding and diffusion on pristine and biaxially-strained 2-dimensional Sc_2C. The latter is metallic and thus might be used as an electrode material for calcium-ion batteries. The pristine 2-dimensional Sc_2C exhibited moderate calcium adsorption, with a relatively low binding energy (0.38eV) on the most stable site. This value increased sharply to -1.94eV and -3.23eV at biaxial compressive strains of 5% and 10%, respectively. The energy barrier to calcium diffusion was only 80meV on pristine 2-dimensional Sc_2C and further decreased to 35meV upon applying a median biaxial compressive strain of 5%. Due to the increased binding of calcium on strained 2-dimensional Sc_2C (figure 6), the maximum stable calcium concentration was also increased. The theoretical specific energy capacity of 2-dimensional Sc_2C under biaxial compressive strain (1051.84mAh/g) was higher than that (878.29mAh/g) for pristine material. The average open-circuit voltages for the two cases were both high, being close to 9.3V for the pristine case and 9.0V after 5% of biaxial compressive strain. This showed that biaxial compressive strain can be used to improve the properties of 2-dimensional materials such as Sc_2C, with regard to the specific-energy capacity and open-circuit voltage.

Table 7. Comparison of the capacity and diffusivity of competing anode materials

Material	Specific Capacity(mAh/g)	Diffusion Barrier(eV)
Ca_7TiC_3	2236	0.13
$Ca_{10}\alpha$-Pn	216.4	0.08
Ca_6-PGDY	2129.1	0.35
Ca_9TiS_2	957.2	0.42
Ca_4Nb_2N	1072	0.051
$Ca_4Ti_3C_2$	319.8	0.11

PGDY: phosphorus-graphdiyne

silicon

The possibility of using the 2-dimensional graphene-like material, silicon carbide nano-sheet, as an anode in rechargeable calcium-ion batteries was investigated[35] by means of density functional theory computations. The electrochemical behaviour during the charging process, the diffusion of calcium ions, the electronic structures and the changes in geometrical structures were investigated. The adsorption energy of calcium on the silicon atom was higher than that for other sites. The maximum theoretical capacity of the 2-dimensional silicon carbide nano-sheet attained 507mAh/g, and this was associated with a slight change in the spacing of the Si-C band. The open-circuit voltage varied with changes in the concentration of adsorbed calcium atoms (figure 7) on the 2-dimensional silicon carbide nano-sheet surface. The open-circuit voltage range of calcium atoms adsorbed on the 2-dimensional silicon carbide nano-sheet indicated that the nano-sheet was suitable as an anodic material. The adsorption energy was -0.731eV for calcium on top of the silicon atom; more than for other sites. There were changes in the open-circuit voltage with increasing numbers of adsorbed calcium atoms. The trend indicated that the 2-dimensional SiC nano-sheet could accommodate up to 9 calcium atoms; corresponding to the observed capacity. The calcium ion migrated on the surface in 2 possible directions. Along path-I the calcium ion diffused mainly between the Si–C bond units.

Materials Research Forum LLC
https://doi.org/10.21741/9781644903490

The energy barrier was 0.23eV. Along path-II the calcium ion diffused mainly on top of the silicon-atom sites, with the maximum diffusion barrier being 0.54eV.

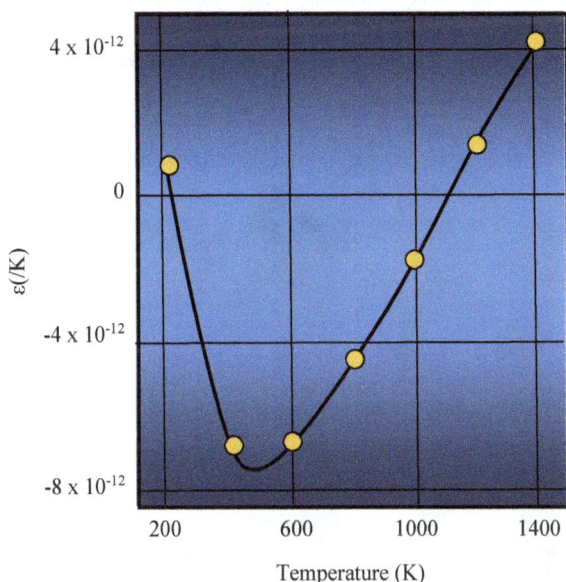

Figure 8. Thermal expansion coefficient of TiC₃
as a function of temperature

titanium

A study was made[36] of the structural, electronic and thermal properties (figures 8 and 9) of TiC$_3$. Calculation of the thermal expansion coefficient over a large part of the temperature range yielded a negative value. The carbon-rich polytype of titanium carbide was considered as an anode material for calcium-ion batteries. The adsorption of Ca^{2+} ions was found to be favourable, with a generous accommodation of guest-atoms and rapid ionic mobility. The total volume expansion for maximum adsorbed Ca^{2+} was 8.2%; a figure which was low when compared with that for other anode materials. Calciation of TiC$_3$ up to the highest Ca^{2+} concentration, Ca$_7$TiC$_3$, led to a theoretical capacity of 2236mAh/g (table 7). With regard to the rate capability, the lowest calculated diffusion barrier was 0.13eV. There was a diffusion coefficient of 0.001cm^2/s along the corresponding pathway, indicating easy movement of calcium ions within the host material. The equilibrium distance of 2.5Å between the host and guest atoms indicated a considerable interaction between them. Among the various adsorption sites, the A-site was concluded to be the most favourable, with an adsorption energy that was lower than that for lithium or sodium. Analysis of the bonding-charge density revealed electron-transfer from Ca^{2+} across all of the selected adsorption sites of TiC_3. All of the configurations of calcium absorption which involved up to 7 Ca^{2+} ions were thermodynamically stable. The carbide retained its metallic structure during the incorporation of guest atoms, thus revealing the stability of its atomic structure.

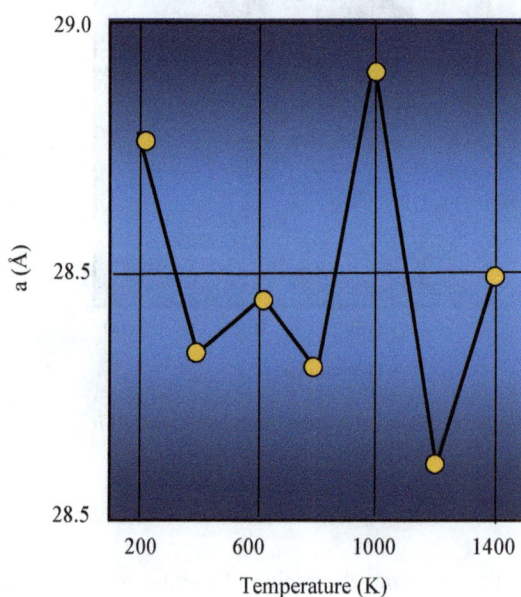

Figure 9. Temperature dependence of the lattice parameter of TiC$_3$

Cyanides

The first electrochemical insertion of Ca^{2+} into the Prussian Blue analogue, $MnFe(CN)_6$, was carried out[37] in non-aqueous solutions of $Ca(CF_3SO_3)_2$ and various solvents at 60C. The kinetics of the Ca^{2+} insertion were studied by means of cyclic voltammetry, and compared with the kinetics of Na^+ intercalation. The process was combined with the use of metallic anodes in order to create two energy storage devices. A calcium anode produced a primary cell which operated at a voltage of around 2.0V. When a magnesium plate was instead used as an anode, negative active material which was associated with $CF_3SO_3^-$ was formed at the surface of the magnesium plate. By exploiting this negative active material, it was possible to create a rechargeable battery by using dual ion-species transport: with Ca^{2+} for the cathode and $CF_3SO_3^-$ for the anode. Ball-milled copper hexacyanoferrate (CuHCF), a Prussian Blue analogue, was investigated with regard to its electrochemical properties as a cathode for aqueous calcium-ion batteries[38]. The ball-milling process did not destroy the crystal structure of the CuHCF, and electrochemical tests showed that prolonged ball-milling improved the charge/discharge capacity during the initial cycle but, following 200 cycles, structural collapse of the CuHCF electrode occurred.

Potassium barium hexacyanoferrate, $K_2BaFe(CN)_6$, was investigated as a cathode material for a reversible Ca^{2+} ion insertion/extraction type of rechargeable battery with a non-aqueous electrolyte[39]. It was shown that the addition of water led to a marked increase in the intercalation and de-intercalation of Ca^{2+} ions, and led to an improved charge/discharge capacity. The improvement in performance was attributed to the formation of solvation spheres, around the intercalating Ca^{2+} ions, which provided screening from the electrostatic charges of the $BaFe(CN)_6$ lattice. A reversible capacity of 55.8mAh/g, and a coulombic efficiency of 93.8%, was found following 30 charge/discharge cycles.

The Prussian Blue analogue, $K_xMFe(CN)_6 \cdot nH_2O$, was investigated as a cathode material by which to insert and extract Ca^{2+} reversibly, using a calcium-based organic electrolyte[40]. The dehydrated $K_xNiFe(CN)_6$ electrode, with a highly electroconductive additive, exhibited reversible capacities of some 50mAh/g and coulombic efficiencies of about 92%. During the first discharge and discharge-charge cycle, it was found that Ca^{2+} was inserted and extracted at interstitial sites of the $K_xNiFe(CN)_6$ without destroying the open framework structure.

Potassium iron hexacyanoferrate, Prussian Blue was itself investigated[41] as a cathode for non-aqueous divalent calcium ion batteries. This material was attractive because of its high specific capacity, non-toxicity, low cost and simplicity of preparation. Charge/discharge cycles were performed at current densities of 23, 45, 90 and 125mA/g and produced reversible specific capacities which ranged 150mAh/g at the 23mA/g current density to above 120mAh/g at the 125mA/g current density. At the time, these were then the highest storage capacities recorded for a divalent calcium-ion cathode

Materials Research Forum LLC
https://doi.org/10.21741/9781644903490

during extended charge/discharge cycling. The performance was comparable to that found for monovalent intercalating ions.

The effect of a change in the hydration number of calcium ions, due to a variation in the electrolyte concentration in a Prussian Blue analogue (CuHCF) electrode was studied[42] for an aqueous electrolyte system. The coordination number of the calcium ions (the number of surrounding nitrate anions in the aqueous electrolyte) gradually increased when the electrolyte concentration was increased from 1.0 to 8.4mol/dm^3. That is, the hydration number of calcium ions gradually decreased. The activation energy for interfacial charge-transfer, during calcium-ion insertion into the CuHCF electrode, was estimated from the temperature dependence of the charge-transfer resistance in aqueous electrolytes (table 8). This activation energy gradually decreased with increasing electrolyte concentration.

Table 8. Activation energy and charge-transfer resistance at 298K for calcium insertion into CuHCF at 0.6V

Concentration(M)	Activation Energy(kJ/mol)	Charge Transfer Resistance(Ω)
1.0	6.14	2.10
1.1	5.87	3.00
2.2	5.84	3.90
4.6	5.60	4.00
8.4	3.80	11.5

Divalent iron-ions were introduced into copper hexacyanoferrate, CuHCF, in order to construct[43] a Prussian Blue cathode material which was rich in Fe^{2+}. This was done by using $K_4Fe(CN)_6$ as the precursor, instead of $K_3Fe(CN)_6$. The Fe^{2+} ions, in a low-spin state, could improve the structural stability of CuHCF during calcium-ion extraction and insertion. The lattice parameter change of CuHCF was only 0.13% during charging and discharging. This was much less than that for CuHCF with Fe^{3+}. X-ray absorption spectroscopy revealed that the charge-compensation of $CuHCF(Fe^{2+})$ was due mainly to a Fe^{2+}/Fe^{3+} redox couple. The octahedral distortion of CuHCF was also effectively suppressed. The $CuHCF(Fe^{2+})$ cathode could thus offer a reversible capacity of 54.5mAh/g at 20mA/g, with a capacity retention of 90.43% following 1000 cycles.

Calcium-Ion Batteries   Materials Research Forum LLC
Materials Research Foundations **175** (2025)   https://doi.org/10.21741/9781644903490

Prussian Blue nano-disk electrodes could markedly increase the lifetime of calcium-based cells by precipitating 20nm-thick nano-disks of Prussian Green from $Fe(NO_3)_3$ and $K_3Fe(CN)_6$, followed by their treatment with NaI[44]. Prussian-Blue cathodes with a polyacrylic-acid/polyaniline binder offered an initial discharge capacity of 77.6mAh/g at 0.1A/g, and retained 91.0% capacity following 700 cycles. The use of polyvinyl fluoride harmed a calcium-based cell because its coulombic efficiency decreased from 94.8% after 120 cycles, to 86.4% after 400 cycles. For a given cathode, a calcium-based cell was much less sensitive to high current densities than was a sodium-based cell. This was attributed to the fact that only half the number of cations was required to move in calcium-based systems, as compared with sodium-based systems. The charge-transfer resistance was thus considerably reduced in calcium-based systems.

Nitrides

Density functional theory calculations were used[45] to evaluate the electrochemical and calcium-storage characteristics of graphyne-like aluminium nitride monolayers as an electrode material for calcium-ion batteries. The changes in internal energy, and the cell voltage of calcium-ion batteries with graphyne-like anodes, were comparable to those of other 2-dimensional nano-materials. The calcium was adsorbed mainly at the centre of a hexagonal or triangular ring of graphyne, with absorption energies of -2.06eV and -0.42eV, respectively. With increasing concentration of calcium atoms on the graphyne, the adsorption energy and the cell voltage decreased. Lower values, 0.15 to 0.32eV, of the diffusion barrier indicated that calcium diffusion in the 2-dimensional nano-sheets was rapid. The graphyne offered a maximum theoretical capacity of about 869.23mAh/g.

On the basis of first principles calculations, a study was made[46] of the use of B_5N_3 as an electrode material for chargeable calcium-ion batteries. Adsorption of calcium atoms reduced the band-gap of B_5N_3 and led to a good electrical conductivity. The B_5N_3 monolayer permitted the double-layer adsorption of calcium atoms on both sides of the monolayer, thus leading to a theoretical capacity of 4463mAh/g for calcium, as compared with capacities of 2231mAh/g for sodium, 1116mAh/g for lithium and 558mAh/g for potassium. The high capacity was attributed to the multiple empty electron-orbitals of the constituent elements of B_5N_3 and a low distance-mismatch which could exhibit good adsorption properties for multivalent atoms. Low energy barriers to diffusion, and acceptable thermal stability, confirmed the suitability of B_5N_3 as an electrode material.

Pristine boron nitride and Stone-Wales defect nitride nano-sheets have been considered[47] as anode materials for calcium-ion batteries. In comparison with pristine BN monolayers, the Stone-Wales nitride offered a higher conductivity, together with calcium adsorption.

A more rapid mobility of calcium ion in Stone-Wales 2-dimensional nano-sheet was associated with a lower diffusion barrier of 0.11eV. The Stone-Wales nitride also offered a suitable open-circuit voltage range for use as anode materials. The maximum theoretical capacity of the Stone-Wales nitride was 1162.66mAh/g. This was some 3 times higher than that (345mAh/g) of pristine BN. Adsorbed calcium atoms were located symmetrically on both sides because the adsorption in that case required a lower formation energy than it did on one side. This continued until excess adsorbed atoms disrupted the original structure. The open-circuit voltage was calculated as a function of the adsorbed calcium-atom concentration (figure 10). The open-circuit voltage decreased at a different rate, for concentrations of less than 0.20, due to an increase in calcium-atom content.

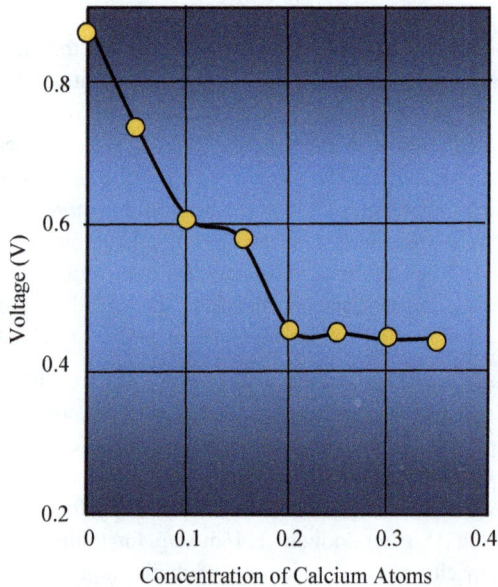

Figure 10. Open-circuit voltage versus calcium adsorption on
Stone-Wales BN nano-sheets

A new group of anode materials for calcium-ion batteries has been identified following the synthesis of carbon nano-tubes and their boron- and nitrogen-doped derivatives, and so BC_2N nano-tubes have been studied[48] as possible anode materials. First-principles computations were used to determine the electrochemical, cycling and adsorption of BC_2N nano-tubes. Nuclear magnetic resonance data revealed the presence of 2 types of non-aromatic hexagonal rings: $B_2C_2N_2$ and $BC4N$. Calcium was adsorbed on $B_2C_2N_2$ and BC_4N with adsorption energies of -47.44 and -28.50kcal/mol, respectively. The specific capacity could be as high as 840mAh/g. The predicted average open-circuit voltage for BC_2N nano-tube was 1.56V, and this was higher than that for other 2-dimensional materials (table 9). These features made BC_2N nano-tubes a suitable anode material for calcium-ion batteries.

Table 9. Open-circuit voltage, diffusion barrier and capacity of BC_2N nano-tubes

Material	OCV(V)	Diffusion Barrier(eV)	Capacity(mAh/g)
BC_2N	1.56	-0.419	840
MoO_2	0.350	-	1256
Ti_3C_2	0.861	-0.270	352
Mo_2C	1.04	-	560
BC_3	0.027	-0.026	783

A study was made of $Ca_{1.5}Ba_{0.5}Si_5O_3N_6$ as a possible calcium solid-state conductor, and its calcium migration was investigated[49] by means of *ab initio* computations and neutron diffraction. This material contains closely-spaced partially occupied calcium sites which facilitate a unique mechanism for calcium migration. Nuclear density maps which were deduced, using the maximum-entropy method, from neutron powder diffraction data constituted evidence for the presence of low-energy 1-dimensional percolation pathways for calcium-ion migration. *Ab initio* molecular dynamics simulations also indicated the existence of a calcium-ion migration barrier of just 400meV when calcium vacancies were present. It was proposed that this led to a so-called vacancy-adjacent concerted ion-migration mechanism. It was concluded that this offered a new understanding of solid-state calcium-ion diffusion and of the design of cation configurations which can exploit interactions between mobile ions so as to promote multivalent ion-conduction in solid-state materials. The key structural features which activated calcium-ion diffusion were unoccupied face-sharing sites across channels or occupied neighbouring sites in channels

The so-called vacancy-adjacent concerted ion migration which was observed was explained in terms of static calcium vacancies which were required in order to create structural features within the activated 1-dimensional channels. The initial calcium-atom hop into one of these unoccupied sites led to a structural feature which created a concerted ion-diffusion mechanism for further hopping.

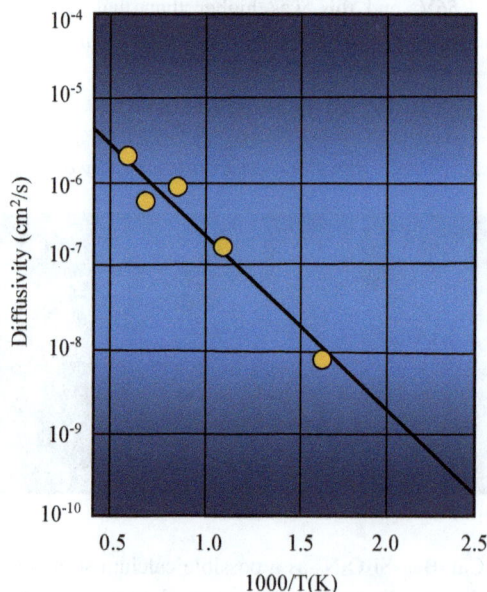

Figure 11. Arrhenius plot of diffusivities in $Ca_{1.5}Ba_{0.5}Si_5O_3N_6$

This concerted ion motion then progressed in its channel until it met an existing static vacancy which annihilated it. A static vacancy might activate calcium-ion diffusion in more than 2 adjacent channels, and so the minimum concentration of calcium vacancies required to sustain percolating calcium-ion conduction could be relatively low. The probability of concerted calcium hopping in the vicinity of the static vacancy was however low, limiting mobility to the channel containing the static vacancy. The unique migration mechanism suggested that interactions between mobile ions played an important role in solid-state ion diffusion. The diffusion of large cations such as Ca^{2+} was expected to be more complicated than that of small cations such as Li^+. Steric effects might have to be considered in the design of large-ion conductors. Such steric effects were here accommodated by the vacancy-adjacent diffusion mechanism, unlike the conventional vacancy-mediated cation diffusion that is usually observed in solid-state ion conductors. The present material was predicted to have an electrochemical stability window of 1.2 to $4.0V_{Ca/Ca2+}$. The results suggested that calcium vacancies were required in order to activate calcium-ion diffusion. In such cases the temperature-dependent diffusivity data indicated Arrhenius behaviour (figure 11). The extrapolated calcium-ion conductivities at room temperature were 5.4 x 10^{-7}S/cm for $1/12V_{Ca}$ and 2.64 x 10^{-6}S/cm for $1/6V_{Ca}$.

Table 10. Comparison of V_2N properties with those of competing electrode materials

Material	Capacity(mAh/g)	OCV(V)	Diffusion Barrier(eV)
V_2N	770.8	1.18	1.14
Ti_3C_2	319.8	1.43	0.118
V_3C_2	539.71	1.4	0.04
V_3C_2/graphene	598.63	1.2	0.15
Mn_2NO_2	343.6	1.02	0.48
Ti_2C	623.72	-	0.33
Ti_2CO_2	487	-	0.55
V_2CO_2	474	-	0.5
Nb_2CO_2	345	-	0.55
$Ti_3C_2O_2$	382	-	0.65
bilayer-Ti_2CO_2	298.04	-	1.6
bilayer-V_2CO_2	288.23	-	1.5
bilayer S-tetrazine	412.46	-	0.6

The storage capacity of C_2N nano-sheet having a so-called holey 2-dimensional structure was studied[50] with regard to lithium, sodium and calcium ions and its use as a battery

Materials Research Forum LLC

https://doi.org/10.21741/9781644903490

anode material. Calculations predicted a capacity of over 500mAh/g, with good mobility of the calcium atoms. All of the above metals interacted strongly with pyridinic nitrogen in the pores and the material offered an initial storage capacity but, due to strong bonding of the first intercalated metals in the pores, they exhibited poor mobility. When the pores were loaded with at least one metal atom, the mobility markedly improved but the trapped metal atoms affected the capacity of the material and made it much lower. This limited the use of the material as an anode material for lithium- and sodium-ion batteries. In the case of calcium, the trapping of some of the calcium atoms had a lesser effect, due to the dual valence, leading to a higher capacity.

Table 11. Adsorption energy and open-circuit voltage of calcium-intercalated V₂N

Concentration(%)	E(eV)	Open-Circuit Voltage(V)
5	-2.13	1.06
10	-2.30	1.15
15	-2.17	1.08
20	-2.12	1.06
25	-2.07	1.03
30	-2.03	1.01
35	-2.18	1.09
40	-2.36	1.18
45	-2.35	1.18

First-principles calculations were used[51] to study the electrochemical energy-storage properties of V_2N as an electrode material for multivalent zinc- and calcium-ion rechargeable batteries. Properties such as the band structure, adsorption energy, diffusion kinetics and open-circuit voltage are investigated by means of density functional theory. The material exhibited metallic properties, plus high structural stability and low diffusion barriers. The calcium- (770.8mAh/g) adsorbed V_2N layers offered better storage capacities than those reported (table 10) for calcium-intercalated V_3C_2|graphene (598.63mAh/g). The open-circuit voltage was calculated by incrementally increasing the intercalation concentration of calcium ions on the nitride monolayers. The ions were intercalated into preferred locations on both sides of the monolayer. The intercalation continued until the adsorption energy was greater than the cohesive energy. This allowed molecules to adhere to the surface without greatly disrupting the original structure. The results showed that, as the calcium concentration increased, the open-circuit voltage ranged from 1.06V at 5% to 1.17V at 50%. The open-circuit voltage remained constant at about 1.1V, indicating a good reversibility of the intercalation. The calculated open-circuit voltage was consistent with the calculated adsorption energy (table 11).

Borophene and Borides

Borophene is a promising electrode material for lithium-, sodium-, magnesium- and Ca-ion batteries due to its anisotropic Dirac properties, high charge-capacity and low-energy barrier to ion diffusion. The synthesis of active stable borophene nevertheless remains difficult with regard to the creation of electrochemical devices. A method has been described[52], for the preparation of borophene aerogels, which begins with hexagonal boron nitride aerogels. Borophene grows between the hexagonal boron nitride layers, using boron-boron bridges as nucleation sites. The borophene forms monolayers which involve sp^2-sp^3 hybridization. The method produces stable borophene aerogels which are compatible with various battery components. The creations included sodium-ion magnesium-ion and calcium-ion batteries which offered capacities of 478mAh/g (0.5C, >300 cycles), 297mAh/g (0.5C, >300 cycles) and 332mAh/g (0.5C, >400 cycles), respectively. A Li-S battery offered a capacity of 1559mAh/g (0.3C, >1000 cycles).

Density functional theory calculations were used[53] to evaluate borophene as an anode material for calcium-ion batteries. The results showed that phosphorus-doping increased the calcium-storage properties of borophene. The maximum adsorption number of calcium atoms in the phosphorus-doped system was 7, with a theoretical capacity of 964mAh/g. Density-of-states analysis showed that the borophene exhibited metallic properties after adsorbing calcium atoms, and this improved the electrical conductivity of the electrode material. Calculation of the energy barrier to diffusion showed that strain had an effect upon calcium diffusion in monolayer borophene. A compressive strain drove calcium diffusion through the borophene. The results indicated that borophene was a suitable electrode material for calcium-ion batteries.

First-principles calculations showed[54] that the 2-dimensional materials, TiB_4 and SrB_8, performed well when used as anodes for calcium-ion and similar batteries. The TiB_4 had a calcium storage capacity of 1176mAh/g and the SrB_8 had a calcium storage capacity of 616mAh/g. They exhibited a good electrical conductivity, whether or not the calcium was adsorbed. The diffusion barriers on both surfaces were low, thus implying a good rate-performance. The average open-circuit voltage was also very low. The lattice parameters of the materials changed very little during the introduction of calcium. In the case of Ti_9B_{36} the change was 0.03%. In the case of Sr_8B_{64} it was 0.004%. These figures promised a good cycling behaviour. The Sr_8B_{64} could accommodate up to 16 calcium atoms, with a stoichiometry of $Ca_{16}Sr_8B_{64}$. The average open-circuit voltage of Ti_9B_{36} upon adsorbing calcium was 0.487V. The average open-circuit voltage of Sr_8B_{64} with calcium was 0.003V. These values were compared with those for other 2-dimensional materials (table 12), showing that the variation in the lattice constants of these borides were very small, even when storing the maximum amount of calcium. It was concluded that the TiB_4 and SrB_8 monolayers are suitable anodes for calcium-ion batteries.

Table 12. Lattice change of host monolayers before and after adsorption of calcium

Material	Lattice Change(%)
$BOCa_{1.51}$	7.22
GeP_3Ca_4	5.90
SrB_8Ca_2	0.004
$Ti_2PS_2Ca_6$	5.74
$Ti_2PSe_2Ca_6$	5.84
$Ti_2PTe_2Ca_6$	5.95
$Ti_2PTe_2Ca_6$	5.95
TiB_4Ca_2	0.03

A density functional theory method was used[55] to study MB_4 monolayers, where M was chromium, molybdenum or tungsten, as anode materials for calcium-ion batteries. The 2-dimensional monolayers were suitable for rechargeable metal-ion batteries because of their layered structure, many accommodating sites and high specific surface area. The monolayers possessed thermally, mechanically and dynamically stable structures (table 13). The energetically favourable adsorption of 6 layers of calcium atoms imparted storage capacities of 3377mAh/g, 2311mAh/g and 1416mAh/g to CrB_4, MoB_4 and WB_4, respectively, together with a low average open-circuit voltage of 0.45V, 0.47V and 0.35V, respectively. Three adsorption sites, termed hollow, bridge and top, were selected (table 14). The greater mobility of calcium ions in the MB_4 monolayers was reflected by activation barriers of 0.67eV, 0.72eV and 0.79eV for CrB_4, MoB_4 and WB_4, respectively.

Table 13. Calculated elastic constants for borides

Material	Lattice Constant(Å)	C_{11}(N/m)	C_{12}(N/m)	C_{66}(N/m)	E(N/m)
CrB_4	8.67	356.26	84.01	136.13	336.45
MoB_4	8.87	402.12	58.05	172.04	391.16
WB_4	8.88	407.76	61.32	173.22	398.53

The metallic nature of the materials was retained following the adsorption of maximum concentrations of calcium ions. Following adsorption of calcium atoms, the adsorption energy steadily decreased as the number of adsorbed calcium ions increased. In the case of the most stable site, the calcium atom transferred a charge of 1.40e to monolayer CrB_4, 1.32e to monolayer MoB_4 and 1.35e to monolayer WB_4 (tables 15 and 16).

Table 14. Calculated adsorption energy and Bader charge of borides

Material	Position	Adsorption Energy(eV)	Bader charge(e)
$Ca_{0.11}CrB_4$	top	-3.20	1.39
$Ca_{0.11}CrB_4$	bridge	-3.93	1.38
$Ca_{0.11}CrB_4$	hollow	-3.94	1.40
$Ca_{0.11}MoB_4$	top	-2.33	1.31
$Ca_{0.11}MoB_4$	bridge	-3.17	1.30
$Ca_{0.11}MoB_4$	hollow	-3.18	1.32
$Ca_{0.11}WB_4$	top	-1.89	1.34
$Ca_{0.11}WB_4$	bridge	-2.69	1.34
$Ca_{0.11}WB_4$	hollow	-2.74	1.35

The greatest distance of the last layer to the substrate was 7.95Å, 7.99Å and 8.32Å for CrB_4, MoB_4 and WB_4, respectively. Covalent and ionic bonds co-existed within the boride monolayers, and this increased great structural stability. The volume expansions of CrB_4, MoB_4 and WB_4 for 6 calcium atoms were 5.94%, 6.16% and 6.31%, respectively.

Table 15. Comparison of properties of MB_4 with those of other materials

Material	Voltage(V)	Diffusion Barrier(eV)	Capacity(mAh/g)
CrB_4	0.45	0.67	3377
MoB_4	0.47	0.72	2311
WB_4	0.35	0.79	1416
SrB_8	0.48	0.67	616
borophene	1.59	0.62	964
2D boron	1.50	0.79	2125
SW-BN	0.89	0.11	1162
C_2N	0.23	0.30	419
$FePS_3$	0.48	1.18	586
TiS_2	1.4	1.16	-
BSi	0.40	1.08	2749

Selenium and Selenides

Selenium was initially investigated as a conversion-type electrode for non-aqueous and aqueous calcium-ion batteries[56]. Selenium offered a specific capacity of 476mAh/g, with an average voltage of $2.2V_{Ca/Ca2+}$ at a current density of 50mA/g and a higher energy-density than that of other cathode materials. Long-term cyclic stability at a current-density of 500mA/g was obtained in non-aqueous electrolytes by encapsulating the selenium in mesoporous carbon. Spectroscopy, and density functional theory, suggested that a multi-step conversion process occurred which involved $CaSe_4$ and Ca_2Se_5 intermediates before achieving the final CaSe phase. This was a very different reaction pathway to that which occurred in other metal-selenium batteries. The use of selenium is extended to an aqueous electrolyte. A 1.1V aqueous calcium-ion battery was created by coupling with a copper-based Prussian Blue electrode.

Table 16. Calculated average adsorption energy
of calcium-adsorbed layer on borides

Boride	Calcium Content	Energy(eV)
CrB_4	1	-1.84
CrB_4	2	-1.36
CrB_4	3	-0.85
CrB_4	4	-0.82
CrB_4	5	-0.36
CrB_4	6	-0.14
MoB_4	1	-1.66
MoB_4	2	-1.49
MoB_4	3	-1.01
MoB_4	4	-0.93
MoB_4	5	-0.42
MoB_4	6	-0.11
WB_4	1	-1.40
WB_4	2	-1.11
WB_4	3	-0.62
WB_4	4	-0.44
WB_4	5	-0.39
WB_4	6	-0.27

Elemental selenium was considered[57] as a possible candidate material for the high-capacity cathode of a calcium-ion battery operating via a conversion mechanism at room temperature. Selenium electrodes offered a reversible specific capacity of 180mAh/g, with a discharge plateau near to $2.0V_{Ca2+/Ca}$ at 100mA/g when using an electrolyte that was based upon calcium tetrakis(hexafluoroisopropyloxy)borate in 1,2-dimethoxyethane and calcium metal. The reversible electrochemical reaction between calcium and selenium was clarified by using synchrotron-based techniques.

Table 17. Geometrical parameters of intrinsic WSe$_2$

Deformation(%)	Bond-Length(Å)
0	2.54
2	2.58
4	2.63
6	2.68
8	2.72
10	2.78

A first-principles method was used[58] to calculate the electronic structure of intrinsic WSe$_2$ and of the calcium-adsorbed WSe$_2$ system during shear deformation. The diffusion barrier of calcium on WSe$_2$ was studied in depth. It was shown that shear deformation could reduce the band-gap of the WSe$_2$ system, and shear deformation could easily lead to a transition from semiconductor to metallic properties. When the monolayer structure was subjected to shear deformation, the bond-lengths also changed (tables 17 and 18). The lengths increased with increasing deformation. Adsorbed atoms affected the outflow of some charges to the substrate, and this changed the structural parameters of the system. In the initial state, the calcium adsorption height was 1.54Å while, when the degree of deformation attained its maximum, the calcium the calcium adsorption height was 1.76Å. The WSe$_2$ structure remained highly symmetrical following deformation, with no appreciable distortion occurring after calcium adsorption.

The binding energy was 2.43eV in the initial state, and remained as high as 2.10eV when the degree of deformation was 10%; indicating that the deformed system remained stable.

This was attributed to extensive charge transfer between calcium and the substrate, which benefited the stability of the adsorption system. With increasing degree of deformation, the binding energy decreased and this was because deformation changed the bond-lengths between the atoms in the adsorption system and thereby affected charge-transfer between the atoms.

Table 18. Geometrical parameters of calcium-adsorbed CaWSe$_2$

Deformation(%)	Bond-Length(Å)	Adsorption Height(Å)	Binding Energy(eV)
0	2.62	1.54	2.43
2	2.69	1.58	2.32
4	2.75	1.62	2.29
6	2.82	1.67	2.18
8	2.87	1.71	2.13
10	2.92	1.76	2.10

The change of binding energy also indicated that deformation could control the open-circuit voltage of the system. By comparing the binding-energies of 3 adsorption-sites, labelled H, B and T, it was shown that deformation had its least effect upon the binding energy (table 19) of the H-site. This was due to a stable charge-transfer between H-site calcium atoms and the substrate. The adsorption of calcium led to a change in the band-structure of WSe$_2$. The contribution of calcium d-electrons led to an increase in the peak ranging from 3 to 6eV. Shear deformation reduced the barrier to calcium diffusion on the WSe$_2$ surface, and there was a tendency to transition to metallic properties. The adsorption of calcium made WSe$_2$ change from semiconductor to metal and promoted the electron density of the WSe$_2$ system. The energy barrier to calcium diffusion on the WSe$_2$ surface was only 0.21eV. The Mulliken charge distribution was calculated in order to clarify further the adsorption mechanism of calcium on the surface (table 20). In the stable state of calcium adsorption, the calcium lost 0.58e. When calcium atoms were adsorbed on the surface of the substrate material, a large quantity of charge was built up between the adsorbed atoms and the substrate surface. Deformation caused calcium to lose less charge, suggesting that deformation weakened the ionic bond behaviour in the calcium-adsorbed WSe$_2$ system. Lost electrons in the adsorbed atoms gathered around the atoms of the substrate and the adsorbed atoms in all systems could bind well to the substrate.

The effect of torsional deformation on the electronic properties of intrinsic WSe_2 and the calcium-adsorbed WSe_2 system was studied[59] using first-principles methods. It was shown that calcium could stably adsorb at the vacancies of the WSe_2 surface in all of the deformation systems; the adsorption energy was highest for the system without deformation. The intrinsic WSe_2 was a semiconductor with a direct band-gap of 1.53eV.

Table 19. Binding energies at various adsorption sites of WSe_2

Deformation(%)	Site	Binding Energy(eV)
0	H	2.43
0	B	1.99
0	T	2.25
2	H	2.32
2	B	1.82
2	T	2.12
4	H	2.29
4	B	1.56
4	T	1.98
6	H	2.18
6	B	1.42
6	T	1.78
8	H	2.13
8	B	1.32
8	T	1.64
10	H	2.10
10	B	1.19
10	T	1.52

Torsional deformation changed the WSe_2 from a direct band-gap semiconductor to an indirect band-gap semiconductor and finally to a material with metallic properties. The adsorption of calcium made the conduction-band of the WSe_2 move downward and increased the number of peaks in the conduction-band region. New density-of-state peaks arose mainly from the contributions of tungsten d-type, selenium p-type and the d orbitals of adsorbed atoms. The calcium transferred most of the valence electrons to the substrate, and torsion changed the amount of transferred charge. Twist deformation reduced the barrier to calcium diffusion on the WSe_2 surface from 0.20 to 0.14eV. These results were a basis for improving the application of WSe_2 to ion batteries.

Table 20. Mulliken charge of calcium adsorption structure of WSe_2

Deformation(%)	Calcium	W(+e)	Se(-e)
0	0.58	0.373	-0.278
2	0.52	0.322	-0.238
4	0.48	0.285	-0.211
6	0.42	0.274	-0.194
8	0.32	0.284	-0.169
10	0.26	0.271	-0.161

Sulphides

copper

Copper sulphide was identified as being a suitable positive electrode material for multivalent-ion batteries. Hierarchical CuS porous nano-cages were prepared[60] by using a simple 1-step room-temperature liquid-phase process and were tested as positive electrode materials for rechargeable batteries of calcium-ion, zinc-ion, iron-ion and aluminium-ion type. The hierarchical CuS porous nano-cages offered promising electrochemical behaviours with regard to multivalent-ion battery systems. The results proved the superiority of the nanostructure in improving the performance of positive electrode materials for multivalent-ion batteries.

niobium

Buckled hexagonal-Nb_2S_2 monolayer has been proposed[61] to be a suitable anode material for calcium-ion batteries, with its inherent metallicity contributing to its electrical conductivity as an electrode. Attention was also paid to the structural and electrical properties of calcium-adsorbed Nb_2S_2 monolayer. Calcium atoms prefer to be intercalated on top of the buckled hexagons of the pristine monolayer, and the metallicity of the 2-dimensional nano-sheet is preserved. Bader charge analysis was used to determine the degree of charge-transfer from the calcium atom to the Nb_2S_2 monolayer. The system offered a storage capacity of 1288.86mAh/g and an energy-barrier to diffusion of 0.76eV. A calculated open-circuit voltage of 0.49V confirmed that the Nb_2S_2 monolayer could be a suitable candidate material for use as an anode.

titanium

The possible use of titanium disulphide in aqueous calcium-ion batteries was investigated[62] by studying the electrochemical redox reactions of calcium ions within TiS_2. The effect of electrolyte concentrations ranging from 1.0 to 8.0mol/dm^3 upon TiS_2 electrode reactions was determined. This showed that TiS_2 exhibited distinct charge and discharge behaviours in various aqueous calcium-ion electrolytes. At higher electrolyte concentrations, the TiS_2 suppressed the hydrogen-generation reaction caused by water decomposition. *In situ* X-ray diffraction data confirmed the intercalation of Ca^{2+} ions between the TiS_2 layers during charging. This was a critical discovery, which confirmed the value of TiS_2 in constructing aqueous calcium-ion batteries. X-ray photo-electron spectroscopy data meanwhile supported the formation of a solid electrolyte interphase on an TiS_2 electrode surface; one which aided the suppression of electrolyte decomposition reactions. The choice of anion which was coordinated with Ca^{2+} ions affected the solid electrolyte interphase formation and cycling behaviour. An investigation was made[63] of the electrochemical intercalation of Ca^{2+} ions into TiS_2 in organic electrolytes at room temperature. This showed that intercalation and de-intercalation was possible by using a 0.1M solution of $Ca(CF_3SO_3)_2$ in propylene carbonate as an electrolyte. Improved charge/discharge capacity, reversibility and hysteresis were observed when a 0.1M solution of $Ca(CF_3SO_3)_2$ in a 1:10 (mol/mol) mixture of propylene carbonate and dimethyl carbonate was used, and this was attributed to modulation of the solvation environment of the Ca^{2+} ions by the use of the relatively low-polar dimethyl carbonate. Structural changes in TiS_2 were caused by the Ca^{2+} intercalation/de-intercalation during charging and discharging.

vanadium

An amorphous vanadium structure was induced by molybdenum-doping[64] and *in situ* electrochemical activation was studied for its use as an anode material for calcium-ion batteries. Doping with molybdenum could destroy the lattice stability of VS_4 material, and increase the flexibility of the structure. Subsequent electrochemical activation converted the material into sulphide and oxide co-dominated composites, MoVSO, which served as an active material for the storage of Ca^{2+} during cycling. This amorphous vanadium structure offered a good rate capability, with discharge capacities of 306.7 and 149.2mAh/g at 5 and 50A/g, respectively, and a life of 2000 cycles with 91.2% capacity-retention. These were higher values than those previously reported for calcium-ion batteries. The material underwent partial phase transition to yield MoVSO. This revealed the calcium storage mechanism of vanadium sulphide in aqueous electrolytes.

Tellurides

The use of transition-metal chalcogenides as anode materials for calcium-ion batteries was investigated[65]. The structural stability, electronic structures and diffusion barriers of bulk $MoTe_2$ and WTe_2 were studied using first-principles density functional theory calculations. A density-of-states analysis revealed a metallic behaviour of the tellurides during calcification. The voltage ranges of Ca_xMoTe_2 and Ca_xWTe_2 were 1.53 to 0.45V and 1.48 to 0.41V, respectively, as x ranged from 0 to 5. The diffusion barrier to Ca^+ through XTe_2 indicated that a compressive strain promoted the diffusion of calcium through XTe_2 (table 21). The lattice parameters were a = b = 3.589Å for $MoTe_2$ and a = b = 3.551Å for WTe_2. A top view of the modelled XTe_2 monolayer envisaged an hexagonal sheet in which the X and Te atoms occupied a sub-lattice. A side-view pictured the X-atoms as forming an intermediate layer which was adjacent to the upper Te layer and the lower Te layer. Each X atom was surrounded by 6 tellurium atoms with a triangular prism coordination. The XTe_2 offered 3 different Ca^+ insertion-sites: the H-site was located above the centre of the hexagon and had the highest energy (figure 12), forming six Te-X-Te bonds with Ca^+. A second stable Ca^+ site (T) was located above the X-atom and formed three Te-X-Te bonds. A third stable Ca^+ site (B) was located between the Te-Te atoms and formed a triangular bond between the 2 tellurium layers. During calcification, Ca^+ was inserted into the H-sites of XTe_2 until the composition reached Ca_2XTe_2. The T-site above the X-atom was then filled with Ca^+. Following the gradual insertion of Ca^+, Ca_4XTe_2 was formed. Following insertion at the first two sites, Ca^+ was inserted between Te-Te atoms so as to fill the B-sites, giving Ca_5XTe_2. The distance between the tellurium atoms and the most distant Ca^+ was termed the adsorption height. During calcification, the adsorption-height of Ca^+ along the c-axis remained enlarged.

Table 21. Lattice parameters and adsorption heights of bulk calciated tellurides

Material	x	a(Å)	b(Å)	Adsorption Height (Å)
Ca_xMoTe_2	0	3.589	3.589	-
Ca_xMoTe_2	1	3.591	3.592	2.488
Ca_xMoTe_2	2	3.605	3.601	2.490
Ca_xMoTe_2	3	3.582	3.582	5.389
Ca_xMoTe_2	4	3.588	3.589	5.397
Ca_xMoTe_2	5	3.711	3.713	5.539
Ca_xWTe_2	0	3.551	3.511	-
Ca_xWTe_2	1	3.572	3.571	2.495
Ca_xWTe_2	2	3.597	3.599	2.499
Ca_xWTe_2	3	3.552	3.556	5.246
Ca_xWTe_2	4	3.544	3.557	5.289
Ca_xWTe_2	5	3.701	3.709	4.966

The structure underwent a slight change in the plane, and this offset the volume expansion caused by the intercalation of calcium ions during charging of a battery. The Ca_xXTe_2 retained structural stability and coordination up to x = 4. When the fraction of Ca^+ increased further, a slight bond distortion occurred in the upper layer of the Te-Mo-Te chain. The tellurium atoms broke their bonds with Mo-Te and began to form bonds with Ca^+ that was located at trigonal B-sites. When the fraction of Ca^+ increased, the same bond-distortion occurred in Te-W-Te chain. In order to determine the effect of uni-axial strain upon Ca^+ diffusion, the diffusion barrier at various strains was calculated (table 22). The diffusion barrier increased slightly under tensile straining and decreased slightly under compressive straining, with 4% of tensile strain increasing the diffusion barrier for $MoTe_2$ from 0.57 to 0.72eV and 4% of compressive strain decreasing it from 0.57 to 0.49eV. The diffusion barrier for WTe_2 underwent analogous changes. These tellurides were concluded to be promising electrode materials for calcium-ion batteries due to their good calcium storage-capacity and low diffusion barriers.

Table 22. Change in calcium-ion diffusion barrier with strain in tellurides

Material	Strain(%)	Barrier(eV)
$MoTe_2$	-2	0.49
$MoTe_2$	-1	0.53
$MoTe_2$	0	0.57
$MoTe_2$	1	0.64
$MoTe_2$	2	0.72
WTe_2	-2	0.36
WTe_2	-1	0.38
WTe_2	0	0.41
WTe_2	1	0.48
WTe_2	2	0.54

First-principles calculations were used[66] to calculate the electronic structure and barrier to diffusion of calcium-adsorbed $MoTe_2$ under various degrees of shear deformation. Both pure $MoTe_2$ and calcium-adsorbed $MoTe_2$ were affected by shear deformation. The pure $MoTe_2$ underwent a transition from direct to indirect band-gap under shear deformation. The adsorption of calcium made $MoTe_2$ change from semiconducting to quasi-metallic. Density-of-states analysis showed that calcium insertion appreciably increased the conduction band part of the adsorption system. The diffusion barrier for calcium through $MoTe_2$ suggested that shear deformation promoted the diffusion of calcium on the surface of $MoTe_2$. Shear deformation could modulate the electronic properties of the $MoTe_2$ system, and this provided a theoretical basis for the use of $MoTe_2$ in ion batteries.

Phosphides

Monolayer BeP_2 exhibits intrinsic metallicity and high conductivity. There exists a direct relationship between the electric conductance and electronic properties of electrode materials, and their capacity and cycling stability. An investigation was thus made of the electronic properties of the original geometry of a BeP_2 monolayer and of its geometry following the adhesion of metal ions. The initial structure had a high carrier concentration, and a density-of-states study showed that the p-orbital of phosphorus was the main reason for the metallic nature.

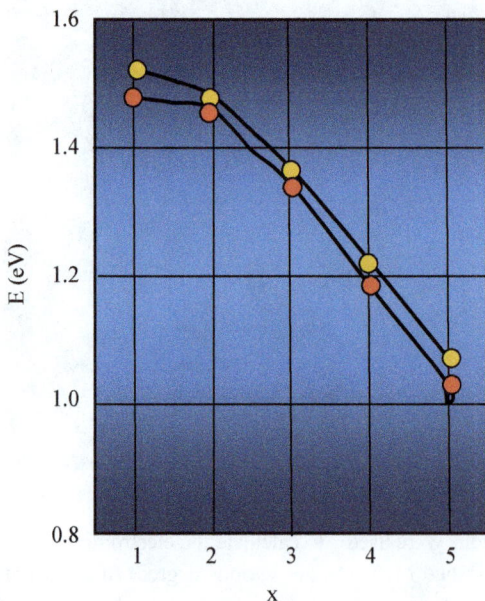

Figure 12. Binding energy of Ca_xXTe_2.
Orange: X = Mo, red: X = W

Electrode materials must offer various adhesion-sites, with low adhesion energies, in order to be used in metal-ion batteries. First-principles density functional theory computations suggested[67] that it would be a suitable electrode material for calcium-ion batteries. The theoretical capacity of a fully calcium phase, Ca_2BeP_2 was 629mAh/g, it had an additional calcium-ion layer which out-performed other known 2-dimensional phases (table 23). Following the full adhesion of metal ions, the BeP_2 had a low open-circuit voltage and a barrier to diffusion of 0.24V. Potential adhesion sites were surveyed for the adhesion of each calcium atom to BeP_2. Sites were chosen according to the symmetrical structure of the phosphide. Three unequal adhesion sites which offered stability were identified following structural relaxation. Those sites which were directly above beryllium and phosphorus were designated as being A1 and A3, respectively. A site between beryllium and phosphorus was designated as A2, but this finally converted to A1. A square hexagonal ring centre, formed by phosphorus and beryllium atoms was designated as being A4 (table 24).

Table 23. Capacity and open-circuit voltage
of various calcium-ion batteries

Material	OCV(V)	Capacity(mAh/g)
BeP_2	0.24	629
$TiPS_2$	0.40	842
Ti_3C_2	0.20	192
VS_2	0.15	257
VO_2	0.75	260

The negative adhesion energies here reflect the tendency of metal atoms, adhered to BeP_2 monolayers, not to form clusters. The A1 site had the lowest absorption energy, and was regarded as being the optimum adhesion site. The adsorption energy for A2 was similar to that for A1. This was attributed to the fact that a calcium atom at A2 adhered at A1 following full structural relaxation. Because of its positive absorption energy, A3 was not a suitable site for adhesion.

A charge-density difference map analysis showed that there was electron-transfer from metal atoms at A1 to phosphorus. The fact that phosphorus ions were more electronegative than calcium ions was consistent with the present results. The number of metal ions which adhered layer-by-layer to BeP_2 monolayers determined the maximum theoretical capacity. The number of layers with adhered calcium ions and the related inter-layer adhesion energies increased gradually as the number of calcium ions increased.

The most favourable adhesion site for the first layer of metal ions was A1, following the adhesion of metal ions at various sites after relaxing the structure. The resultant average inter-layer adsorption energies remained negative when metal ions occupied the A1 sites on both sides; thus allowing for the possibility of further adhesion. The efficient adhesion site for the second layer was A2. The metal ions adhered at suitable adhesion sites, and the overall structure was optimized until full convergence of the energy was satisfied without deformation.

Table 24. Adsorption energy and diffusion barrier
for single calcium ions on BeP$_2$

Site	Absorption Energy(eV)	Diffusion Barrier(eV)
A1	-0.809	0.086
A2	-0.796	0.121
A3	0.266	-
A4	-2.344	0.095

Another site was then selected for the adhesion of further metal ions. A negative total energy difference or adhesion energy showed that adhesion of this layer of ions was capable of decreasing the total energy and increasing stability. This encouraged the adhesion of further ions. When the average inter-layer absorption energy became positive, the theoretical maximum of the adhesion layers could be predicted. When the open-circuit voltage of the electrodes was low (table 25), the net cell voltage was high.

Phosphates

Two polyanionic phosphate materials were identified[68] as being suitable high-voltage cathodes for calcium-ion batteries at room temperature. Thus $NaV_2(PO_4)_3$ electrodes were found to intercalate reversibly 0.6mol of Ca^{2+} (81mAh/g) near to $3.2V_{Ca2+/Ca}$ with a stable cycling performance at a current density of 3.5mA/g. Olivine-structured $FePO_4$ reversibly intercalated 0.2mol of Ca^{2+} (72mAh/g) near to $2.9V_{Ca2+/Ca}$ at a current density of 7.5mA/g during the first cycle. The structural, electronic and compositional changes were consistent with reversible Ca^{2+} intercalation.

Polyanionic K_xVPO_4F, where x was essentially equal to zero, was suggested[69] to be a suitable high-voltage ultra-stable cathode material. It exhibits an acceptable calcium-storage capacity of 75mAh/g at 10mA/g and an appreciable capacity-retention of 84.2% over 1000 cycles. The average working voltage is $3.85V_{Ca2+/Ca}$. Small volume changes and hopping-diffusion barriers lead to a marked stability and to high-power capabilities, respectively. The distribution of calcium ions into polyanionic frameworks with high spatial separation lowers the Ca^{2+}–Ca^{2+} repulsive force and thereby increases the calcium-migration kinetics. The high voltage was attributed to an inductive effect arising from the largely electronegative fluorine. When combined with a calcium-metal anode and a suitable electrolyte, a cell offered an energy density of about 300Wh/kg.

Table 25. Open-circuit voltage of Ca_xBeP_2 monolayers

x	Open-Circuit Voltage(V)
0.50	0.53
1	0.39
1.50	0.27
2.0	0.24

Polyanionic $K_3V_2(PO_4)_3$/C was studied[70] as a cathode material for aqueous calcium-ion batteries. Due to the robust structure of the polyanionic material and the wide electrochemical window for water-in-salt electrolytes, the phosphate offered a working voltage of $3.74V_{Ca2+/Ca}$, with a specific capacity of 102.4mAh/g and a lifetime of 6000 cycles at 500mA/g. The calcium-storage mechanism involved the coexistence of a solid solution and a two-phase reaction. An aqueous calcium-ion cell which was based upon an organic anode and a $K_3V_2(PO_4)_3$/C cathode was constructed. It exhibited good stability over 200 cycles and a specific capacity of 80.2mAh/g.

A proposed intercalation host exhibited a capacity-retention of 90% over 500 cycles, together with a power capability at about $3.2V_{Ca/Ca2+}$, when incorporated into a calcium-ion battery[71]. The cathode material was based upon $Na_{0.5}VPO_{4.8}F_{0.7}$ and could reversibly accommodate a large number of Ca^{2+} ions so as to form $Ca_xNa_{0.5}VPO_{4.8}F_{0.7}$, where x ranged from 0 to 0.5, without any appreciable structural degradation. The strong structure led to a volume-change of only 1.4%, plus a very low barrier to Ca^{2+} diffusion.

The highly reversible calcium-intercalation ability of $NaV_2(PO_4)_3$ makes it a candidate cathode material for non-aqueous calcium-ion batteries. The capacity, voltage and cyclability were 90mAh/g and about 3.4V at 11.7mA/g and 75C or 70mAh/g and about 3.2V at 5.85mA/g and 25C[72].

The cathode material, $VOPO_4 \cdot 2H_2O$, when incorporated into a calcium-ion battery led to a discharge capacity of 100.6mAh/g, good cycling stability at up to 200 cycles and a rate performance of 42.7mAh/g at 200mA/g[73]. The calcium-ion-storage mechanism involved a single-phase reaction which was based upon asymmetrical Ca^{2+} insertion and de-insertion.

Table 26. Calcium-ion storage materials for cathodes

Cathode	Electrolyte	CD	Capacity$	Cycles
$CaCo_2O_4$	1M $Ca(ClO_4)_2 \cdot 4H_2O$/CAN	50A	94-78	30
MoO_3	0.5M $Ca(TFSI)_2$ in CAN	$50A/cm^2$	186-116	1
$VOPO_4 \cdot 2H_2O$	0.8M $Ca(TFSI)_2$ **	20mA/g	70-50	200
$Mg_{0.25}V_2O_5 \cdot H_2O$	$Ca(TFSI)_2$*	100mA/g	70.2-60	500
Ca_xMoO_3	0.5M $Ca(TFSI)_2$ in DME	2mA/g	140-80	12
$NH_4V_4O_{10}$	$Ca(ClO_4)_2 \cdot xH_2O$ in CAN	100mA/g	150-150	100
$KNiFe(CN)_6$	0.5M $Ca(TFSI)_2$ in CAN	25A/cm	50-40	12
$K_2BaFe(CN)_6$	1M $Ca(ClO_4)_2$***	12.5mA/g	15-55.8	30
$NaMnFe(CN)_6$	0.2M $Ca(PF_6)_2$ in EC:PC	35mA/g	160-50	35
$KFe[Fe(CN)_6$	1M $Ca(ClO_4)_2$ in CAN	125mA/g	120-103	80
$Ca_3Co_2O_6$	0.45M $Ca(BF_4)_2$ in EC:PC	0.005C	180-75	1
NASICON-$NaV_2(PO_4)_3$	1M $Ca(TFSI)_2$ in CAN	3.5mA/g	80-83	40
NASICON-$NaV_2(PO_4)_3$	0.5M $Ca(BF_4)_2$ in EC:PC	5.85mA/g	70–70	30
$NaFePO_4F$	0.2M $Ca(PF_6)_2$ in EC:PC	10mA/g	95-80	50
$Ti_2O(PO_4)_2(H_2O)$	0.8M $Ca(BF_4)_2$ in EC:PC	20mA/g	101-85	200
$Ti_2O(PO_4)_2(H_2O)$	0.8M $Ca(BF_4)_2$ in EC:PC	50mA/g	61-58	1500

$initial-final (mAh/g), CD: current density, *in quaternary ester carbonates, **in EC:PC:EMC:DMC, ***in ACN+17vol%H_2O, ACN: acetonitrile; EMC: ethylmethyl carbonate; DMC: dimethyl carbonate; DME: dimethoxyethane.

The material, $Ti_2O(PO_4)_2\bullet H_2O$, was proposed[74] to be a Ca^{2+}-insertion electrode for calcium-ion batteries (table 26). The hydrothermally prepared material could reversibly store some $0.51Ca^{2+}$, giving 85mAh/g at about $2.6V_{Ca/Ca2+}$ at room temperature. The material offered a very long cyclability by retaining some 95% of the initial capacity after 1500 charge/discharge cycles (figure 13).

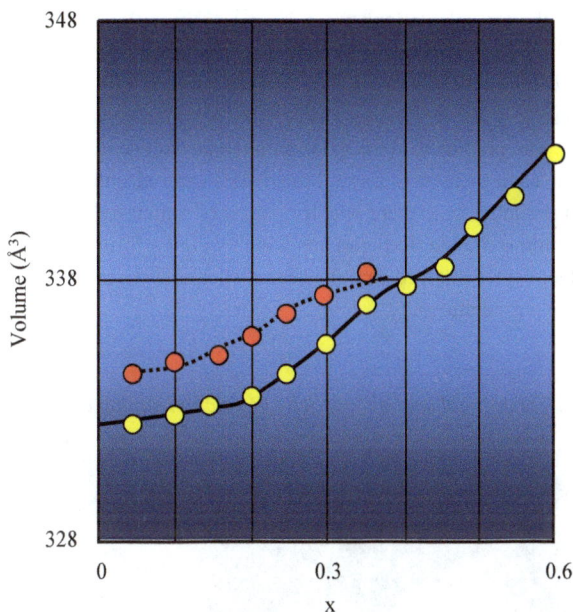

Figure 13. Evolution of lattice volume of $Ca_xTi_2O(PO_4)_2(H_2O)$ during charge (red) and discharge (yellow)

Structural and compositional studies confirmed that reversible Ca^{2+} insertion was associated with Ti^{4+}/Ti^{3+} redox. The electrochemical insertion/extraction of Ca^{2+} ions in $Ti_2O(PO_4)_2\bullet H_2O$ was associated with a small dimensional change and no transformation of the structure. This was suggested to explain the high stability. Bond valence site energy and density functional theory calculations showed that, unlike the reversible Li^+ insertion at about 1.7V in isostructural $M_{0.5}TiO(PO_4)$, where M was Cu^{2+}, Mg^{2+}, Co^{2+}, Ni^{2+} or Fe^{2+}, there was a marked increase to 2.6V in $Ti_2O(PO_4)_2\bullet H_2O$ that was related to an unique Ca^{2+} location and migration path which was attributed to various orientations of the PO_4 tetrahedra and TiO_6 octahedra. A calcium-ion cell was constructed with a configuration of $K_{metal}|Ti_2O(PO_4)_2\bullet H_2O$ by using K^+-containing hybrid electrolytes. This revealed that $Ti_2O(PO_4)_2\bullet H_2O$ is a suitable cathode for calcium-ion batteries.

Silicon and Related Materials

Density functional theory calculations were used[75] to investigate the electrochemical characteristics of silicon-based anodes in calcium-ion batteries (table 27). The calculated average voltage of calcium-alloying with face-centred cubic silicon, to form intermetallic Ca_xSi phases, where x = 0.5 to 2, was 0.4V, with a volume variation of 306%. De-calciation of $CaSi_2$ was predicted to occur at an average voltage of between 0.57V and 1.2V. In the former case there was formation of face-centred cubic silicon and a 65% volume change. In the latter case there was formation of a metastable de-inserted silicon phase and a 29% volume change. Experiments which were performed using conventional alkyl carbonate electrolytes showed that the electrochemical de-calciation of $CaSi_2$ was possible at moderate temperatures.

Table 27. Theoretical specific capacity of calcium-silicon compounds

Compound	Specific Capacity(mAh/g)	Group	a(Å)	b(Å)	c(Å)
silicon	3818	Fd-3m	5.468	-	-
de-inserted silicon	-	R-3m	3.86	-	24.34
$CaSi_2$	557	R-3m	3.84	-	31.68
CaSi	787	Cmcm	4.54	10.74	3.90
Ca_5Si_3	852	I4/mcm	7.63	-	14.79
Ca_2Si	991	Pnma	7.60	4.82	9.04

Theoretical designs for porous 3-dimensional allotropes (ortho- and mono-) of silicene (table 28) had stable structures, high porosities, were conductive and exhibited a high uptake of monovalent and multivalent ions[76]. Due to their porous inner skeletons, low density and inherent conductivity, both allotropes offered great promise as candidate anode materials. The porous silicene structures offered possible capacities of 718 to 1117mAh/g (table 29), together with high average potentials and very low volume expansions during the charging and discharging of calcium ions. Both of the structures had low densities and a conductivity arising from silicon p-orbitals. Their porous structures and the covalent bonding of the silicon atoms furnished many active adsorption-sites and high charge-transfer for calcium ions. The activation energies ranged from 0.36 to 0.48eV for calcium ions. The average open-circuit voltage was 0.86V for the ortho-silicene and 1.76V for the monosilicene

Table 28. Lattice parameters and density of 3-dimensional silicenes

Silicene	Space-Group	a(Å)	b(Å)	c(Å)	α(°)	β(°)	γ(°)	ρ(g/cm³)
ortho-silicene	PMMM	17.65	17.65	3.89	90	90	90	0.93
monosilicene	C2/M	20.09	3.86	6.66	90	89.62	90	1.43

Silaborane clusters, $SiB_{n-1}H_n^-$ with n = 5 to 15, were studied[77] by means of density functional theory. The vertical detachment energy, electrochemical stability window and binding energy of the silaboranes were deduced at the same level of density functional theory. The vertical detachment energy was a measure of the stability of an anion, and was the required to eliminate an electron from an anion with no structural relaxation. A greater value indicated a higher stability of the anion.

Table 29. Comparison of 3-dimensional silicenes with competing anode materials

Material	Capacity(mAh/g)	Diffusion Barrier(eV)	Open-Circuit Voltage(V)
3D-ortho-silicene	1117	0.48	0.86
3D-monosilicene	718	0.32	1.76
2D-TiS₂	957	0.42	-
2D-Ti₃C₂	320	0.11	1.43
3D-PGDY	2129	0.35	0.35

PGDY: phosphorus-graphdiyne

The -NCS and -CF$_3$ substituted SiB$_{11}$ ions were best, according to density functional theory. *Ab initio* molecular dynamics studies clarified the interactions between silaborane-based electrolytes and calcium, and highlighted the decomposition of -NCS and -CF$_3$ substituted SiB$_1$ on calcium anodes. These studies showed that -CH$_3$ substituted silaborane-based calcium, Ca(SiB$_{11}$H$_{11}$CH$_3$)$_2$, was a suitable electrolyte for calcium-ion batteries.

Table 30. Vertical detachment energy of carboranes, CB$_{n-1}$H$_n^-$

n	Vertical Detachment Energy(eV)
5	3.42
6	3.92
7	4.24
8	3.88
9	4.28
10	5.70
11	4.69
12	6.05
13	5.28
14	5.10
15	5.30

Candidate anions having higher values could can act as efficient electrolytes. Values were predicted for carboranes and silaboranes (tables 30 and 31). The electrochemical stability window of an anion indicated the potential window within which the anion was neither oxidized nor reduced. Anions having higher values were more suitable electrolytes (tables 32 and 33). A method which was based upon molecular electrostatic potential surface analysis was used to identify the most suitable binding site for calcium ions on the clusters. The density functional theory results showed that the SiB$_{11}$ cluster was better than other candidates. The effects of substitution with -CH$_3$, -NCS, -CF$_3$, -F or -Cl were calculated.

Table 31. Vertical detachment energy of silaboranes, SiB$_{n-1}$H$_n^-$

n	Vertical Detachment Energy(eV)
5	3.22
6	4.20
7	4.57
8	3.51
9	4.38
10	4.59
11	4.54
12	6.01
13	5.03
14	5.40
15	4.63

Oxides

boron

First-principles calculations were used[78] to determine the calcium capacity of B$_3$O$_3$ monolayers. Calcium atoms could be adsorbed on a B$_3$O$_3$ surface, with the most stable location being the top of the pore-centre of a B$_3$O$_3$ monolayer. The binding energy of B$_3$O$_3$ monolayer was relatively high for calcium atoms. The calcium atoms diffused easily on the B$_3$O$_3$, and the lowest diffusion barrier was 65meV. The B$_3$O$_3$ monolayer-based nanostructures exhibited a capacity of 616.05mAh/g, indicating that the charge-discharge rate was relatively rapid. The complete adsorption of calcium on the B$_3$O$_3$ monolayer led to an open-circuit voltage of about 0.142V, indicating that the operating voltage for full-cells would be high. The ability to bind calcium is an important factor with regard to the choice of an electrodes for calcium storage. Three possible sites (table 34) were investigated for adsorption on the monolayers, among a total of 7. Site 1 was atop the B-B bonds. Site 6 was in the pore centres of the monolayer. Calcium adsorption at sites 1, 6 and 7 sites was spontaneous. Adsorbed calcium atoms on sites 1, 6 and 7 remained there, apart from changes in vertical height. All of these 3 adsorption sites were stable and useful for calcium adsorption. The adsorption energies were negative, indicating that calcium adsorption was exothermic for sites 6 and 7 and that they were both stable. The most stable adsorption site for a single calcium was 6. In comparison with site 7, site 6 was more energetically favourable. There was a relatively great charge-transfer for both sites 6 and 7. Only site 6 was considered in further detail.

Table 32. Electrochemical stability window of carboranes, $CB_{n-1}H_n^-$

n	Electrochemical Stability Window(V)
5	8.15
6	8.48
7	8.63
8	8.18
9	8.52
10	9.26
11	8.76
12	10.10
13	9.17
14	9.02
15	8.96

calcium

Composites of Co_3O_4 and Ca_4O_4 were prepared[79] from a bimetallic-organic framework by using simple wet chemical precipitation and annealing. This nanomaterial was then studied as a possible electrode material in various calcium-ion based organic electrolytes.

Galvanostatic charge-discharge tests showed that the composites offered a maximum specific discharge capacity of 214.31mAh/g in 0.5M $Ca(ClO_4)_2$/acetonitrile electrolyte with an average discharge voltage of $2.4V_{Ca/Ca2+}$ after 10 cycles at 500mA/g. Discharge capacities of 169.58mAh/g and 168.81mAh/g were found when using 0.5M $Ca(NO_3)_2$ in 1:1(v/v) ethylene carbonate/propylene carbonate and 0.5M $Ca(ClO_4)_2$/ECPC electrolytes with average discharge voltages of 1.9V and $1.65V_{Ca/Ca2+}$, respectively.

Table 33. Electrochemical stability window of silaboranes, $SiB_{n-1}H_n^-$

n	Electrochemical Stability Window(V)
5	7.80
6	8.31
7	8.85
8	7.73
9	8.65
10	8.70
11	8.61
12	10.13
13	8.94
14	9.40
15	7.97

cobalt

A calcium-ion battery was described[80] which employed P2/m $CaCo_2O_4$ as the positive electrode, Pmmn V_2O_5 as the negative electrode and calcium perchlorate in acetonitrile as the electrolyte solution. Electrochemical tests showed that the calcium ions could be initially removed from $CaCo_2O_4$ and then re-intercalated. The material was concluded to be a suitable candidate for the development of non-aqueous calcium-ion batteries.

iron

There is an interest in identifying suitable calcium-containing oxides for use as electrode materials in calcium-ion batteries. An atomic-level computational investigation of ionic defects and calcium-ion diffusion in calcium-bearing oxides is important for predicting their suitability for use in calcium-ion batteries. Atomistic simulations were used[81] to determine the energetics of defects, dopants and calcium-ion diffusion in $Ca_3Fe_2Si_3O_{12}$. The calculations suggested that the Ca/Fe antisite defect was the most important intrinsic defect that caused significant disorder and which would be sensitive to the preparation conditions. The diffusion of Ca^{2+} ions within the $Ca_3Fe_2Si_3O_{12}$ was 3-dimensional, and the activation energy for migration was 2.63eV; implying that slow ionic conductivity was involved. The most responsible isovalent defects were deemed to be Mn^{2+}, Sc^{3+} and Ge^{4+} on calcium, iron and silicon, respectively. The presence of extra calcium was concluded to increase the capacity and diffusion of calcium. It was found that Al^{3+} and Mn^{2+} were candidate dopants for silicon and iron sites, respectively, and there was a reduction in the activation energies.

Table 34. Characteristics of a calcium cap on a B_3O_3 monolayer surface

Site	E(eV)	h(Å)	d(Å)	Q(e)
1	-0.017	2.76	3.912	0.360
6	-0.644	2.31	3.604	0.238
7	-0.354	2.54	3.623	0.339

E: adsorption energy, h: vertical height of calcium, d: distance between a calcium atom and the 3 closest neighbours, Q: charge transfer

manganese

The electrochemical behaviour of marokite ($CaMn_2O_4$) is such that experimental attempts to de-insert calcium ions from this compound failed. First-principles calculations indicated[82] that marokite could support reversible calcium de-insertion reactions to the extent that semi-decalciation was predicted to occur at an average voltage of 3.7V, in conjunction with a volume change of 6%. On the other hand, the expected barriers (1eV) to calcium diffusion were too high and explained the observed difficulty of de-inserting calcium from the marokite structure. The theoretical investigation was extended to $CaMn_2O_4$ polymorphs of spinel and $CaFe_2O_4$ structural type. Complete calcium extraction from those polymorphs was predicted to occur at an average voltage of 3.1V, combined with a volume change of about 20%. There remained the potential interest of the spinel as a cathode for calcium-ion batteries, but calculations indicated that its synthesis was not feasible.

Transition-metal oxide post-spinels are expected to possess crystal structures which can provide low migration-barriers, high voltages and the easy transport of calcium ions, thus making them suitable cathodes for calcium-ion batteries. Calcium manganese oxide in the post-spinel form was studied[83] as an intercalation cathode for such batteries. The $CaMn_2O_4$ was first synthesized using solid-state methods, and the desired phase was produced. The redox activity of the cathode was investigated via cyclic voltammetry and galvanostatic cycling. This detected oxidation potentials at 0.2 and 0.5V, and a broad insertion potential at -1.5V. The material could cycle at a capacity of 52mAh/g. It was concluded that $CaMn_2O_4$ is a promising cathode for calcium-ion batteries.

Birnessite-MnO_2 is a promising cathode because of its large interlayer distance and consequently easy Ca^{2+} intercalation. The birnessite-type material, $K_{0.31}MnO_2 \cdot 0.25H_2O$, had a layered structure with interlayer distances of about 7Å. A reversible electrochemical reaction was detected[84] in cyclic voltammograms and galvanostatic cycles. The initial specific discharge capacity was 153mAh/g at 25.8mA/g (0.1C) in a 1M $Ca(NO_3)_2$ aqueous electrolyte, with an average discharge voltage of $2.8V_{Ca/Ca2+}$. X-ray diffraction, transmission electron microscopy and elemental analyses confirmed that Ca^{2+} transport was largely responsible for the electrochemical reaction. The reaction mechanism could be described in terms of a combined surface-limited capacitance and a diffusion-controlled intercalation.

Recalling the fact that the large (1.00Å) size of Ca^{2+} ions limits its ability to intercalate a host structure, a study was made[85] of hierarchical calcium-birnessite with its ultra-thin and intertwined-nanosheet structure and Mn_3O_4 nano-wall arrays. Samples were prepared by using electro-conversion method, and exhibited a rapid surface Faraday reaction as well as favourable Ca^{2+} storage kinetics. Electro-converted calcium-birnessite electrodes offered a specific capacity of 175mAh/g at 0.1A/g, a rate performance of 89mAh/g at 2A/g and extended cycling stability with 83.2% retention after 2000 cycles at 1A/g. The reaction mechanism was dominated by surface-limited capacitance. A 1.9V cell was constructed which offered a reversible capacity of 42.1mAh/g at 0.2A/g and an energy density of 30.56Wh/kg at a power density of 143.4W/kg. It exhibited high cycling stability with essentially no capacity-decay after 1500 cycles at 1A/g.

Cobalt-doped K-birnessite $K_{0.11}Co_{0.02}Mn_{0.98}O_2 \cdot 1.4H_2O$ was investigated[86]. Following cobalt doping, the Mn–O ionic bond was shorter and the band energy was reduced, thus inhibiting the Jahn-Teller effect and improving the thermodynamic stability. The Ca^{2+} could restrain the irreversible transition of birnessite from layered to spinel. The cathode offered a rate performance of 181.2mAh/g at 200mA/g and structural stability to 1000 cycles at 2A/g with 90.4% capacity-retention. A battery was constructed by using the present cathode and a polyimide anode. It offered a specific capacity of 51.9mAh/g at 200mA/g, with good cycling stability.

Four MnO_2 polymorphs (α-, β-, γ-, δ-phase) were considered[87] as cathode materials. The δ-MnO_2 exhibited the best electrochemical performance, due mainly to the higher calcium-ion diffusivity of δ-MnO_2. The observed rate capability out-performed that of other oxide cathode materials for calcium-ion batteries. The calcium-ion storage mechanism of δ-MnO_2 was based upon a reversible order–disorder transformation. The capacity-decay mechanism of δ-MnO_2 during cycling involved the dissolution of manganese.

molybdenum

Layer-structured orthorhombic α-MoO_3 was investigated as an electrode for calcium-ion batteries with a calcium-based organic electrolyte[88]. The ground α-MoO_3, with a low-reactivity carbon additive, offered a capacity of some 120mAh/g per one-electron reaction. A large irreversible capacity was observed due to electrolyte decomposition during the initial (Ca^{2+}-insertion) discharge. X-ray diffraction, following the initial discharge and charge, revealed a decreased intensity of the 0k0 reflections with increased peak-widths as compared with the patterns of as-prepared material. This suggested that structural damage had occurred in the layered structures. Other observations of discharged, and discharged–charged electrodes, showed that Ca^{2+} ions were inserted into, and extracted, from the structures via a two-phase reaction.

Building upon the knowledge that layered α-MoO_3 can intercalate hydrated calcium ions in an aqueous electrolyte, and that such intercalation could increase the interlayer spacing, it was found that α-MoO_3 was electrochemically active in a calcium cell with a non-aqueous electrolyte[89]. The use of 1,2-dimethoxyethane as an electrolyte solution and calcium metal as the negative electrode was demonstrated. In non-aqueous solutions, non-solvated calcium ions were intercalated into the molybdite structure. The layer structure was preserved but the space-group changed. The layered structure was preserved during the electrochemical intercalation of unsolvated calcium and the perovskite-type $CaMoO_3$ structure was not formed. The lattice parameter perpendicular to a slab increased from 13.85 to 14.07Å during the first stages of intercalation, and this modest increase (table 35) could be optimum for offering good electrochemical cycling. A model for calcium intercalation in the interlayer space was optimized by using theoretical calculations which were based upon density functional theory. The experimentally measured reversible capacity was 80 to 100mAh/g and the average voltage was about $1.3V_{Ca}$. It was concluded that improvements in electrolyte composition, particle size and morphology could make molybdite a suitable electrode for rechargeable calcium-ion batteries. The small change in interlayer spacing could be advantageous with regard to structural stability and good electrochemical cycling. It was concluded that the composition, Ca_xMoO_3, should be limited to around x = 0.3 in order to achieve reversible electrochemical cycling. This was attributed to the redox instability of the electrolyte solutions and to the metastable nature of the calcium-rich compositions. Some of the capacity was attributed to surface contributions and conversion reactions. It was noted that the slow diffusion of calcium ions, the existence of side-reactions and the occurrence of a competing conversion reaction could impede progress. The electrolyte composition might be further optimized in order to increase stability at the cathode and increase capacity, but the slow diffusion of calcium and the competing conversion reaction could be drawbacks.

Table 35. Unit-cell parameters of Ca_xMoO_3

x	a(Å)	b(Å)	c(Å)
0	3.96156	13.8550	3.69642
0.05	3.8849	14.073	3.7306
0.1	3.8843	14.071	3.7304
0.2	3.8919	14.072	3.7345

Only a few calcium-insertion electrode materials have been reported, most of which offer only a low capacity or poor cyclability in non-aqueous electrolytes. On the other hand the calcium molybdenum bronze, $Ca_{0.13}MoO_3(H_2O)_{0.41}$, was shown[90] to be a potential calcium-ion battery room-temperature cathode material which offered a reversible discharge capacity of 192mAh/g at 86mA/g with an average voltage of $2.4V_{Ca/Ca2+}$, plus excellent cyclability.

Although α-MoO_3 can reversibly intercalate calcium ions there is a problem with its limited electrochemical activity, and the associated reaction mechanisms remain unclear. Further investigation[91] of the calcium insertion improved the reaction kinetics and clarified the storage mechanism, leading to α-MoO_3 electrodes which offered a specific capacity of 165mAh/g at $2.7V_{Ca2+/Ca}$, together with stable long-term cycling and a good rate performance at room temperature.

The electrochemical Ca^{2+} storage ability of electrodeposited molybdenum oxide has been studied[92]. The d-electron concentration of molybdenum and the presence of structural water in the oxide play an important part in Ca^{2+} storage. The Mo-O-H components in MoO_x can prevent dissolution in a weakly alkaline $CaCl_2$ electrolyte, with a MoO_x electrode offering a discharge capacity of 106mAh/g at 0.1A/g (mass loading of $10mg/cm^2$) and a cycle life of 30000 cycles at 3.0A/g. The MoO_x underwent pure Ca^{2+} insertion and de-insertion during the charge/discharge processes. An aqueous calcium-ion full cell, which combined a MoO_x anode with a copper hexacyanoferrate cathode, offered an areal energy density of $0.96mWh/cm^2$ (42.7Wh/kg) and a long cycle life.

titanium

A layered $Na_2Ti_3O_7$ structure was proposed[93] as an anode material for non-aqueous calcium-ion batteries. Such anodes offered a discharge capacity of 165mAh/g at 100mA/g and 80% capacity-retention after 2000 cycles at 500mA/g. This material out-performed previous intercalation-type anode materials. The $Na_2Ti_3O_7$ changed to layered $Ca^{VII}Na^{IX}Ti_3O_7$ upon intercalation with Ca^{2+} and extraction of Na^+ during the initial discharge. The $Ca^{VII}Na^{IX}Ti_3O_7$ then underwent reversible insertion/extraction of Ca^{2+} during further cycling. Density functional theory calculations indicated that $Na_2Ti_3O_7$ offered a rapid 2-dimensional diffusion pathway for Ca^{2+}.

First-principles calculations were used[94] to show that P-type layered calcium transition metal oxide materials, CaM_2O_4, where M was titanium, vanadium, chromium, manganese, iron, cobalt or nickel exhibit good battery-relevant properties such as thermodynamic stability, average voltage, energy density, ionic mobility and electronic structure. The thermodynamic stability of the charged phase, and the transition-metal redox activity, were sensitive to the nature of the metal; with $CaCo_2O_4$ offering the best balance of those properties. It was also possible to use mixtures of substituted transition metals.

vanadium

On the basis of pre-insertion chemistry, the cycle life of vanadium oxides was improved[95] by the use of integrated electrode and electrolyte choices. By using a tailored calcium electrolyte, a free-standing $(NH_4)_2V_6O_{16} \bullet 1.35H_2O$, graphene oxide and carbon nano-tube composite cathode could offer a capacity of 305mAh/g and a lifetime of 10000 cycles. A calcium-ion hybrid capacitor full cell was created which offered a capacity of

62.8mAh/g. The calcium storage mechanism of the composite cathode involved a 2-phase reaction with the exchange of NH_4^+ and Ca^{2+} during cycling. Lattice self-regulation of the V-O layers occurred and layered vanadium oxides with Ca^{2+} pillars which formed by ion-exchange exhibited a higher capacity.

A design principle was proposed for the preparation of high-solvation electrolytes which offer ultra-stable calcium-ion storage[96]. The decomposition of bis (trifluoromethanesulfonyl) azanide (TFSI) ions and the formation of a CaF_2-rich cathode electrolyte interface with Ca^{2+} insulation can be suppressed in such electrolytes. The electrolyte, $Na_2V_6O_{16}\cdot 2.9H_2O$, offered a discharge capacity of 240.7mAh/g at 20mA/g and a lifetime of 60000 cycles at 1000mA/g. A 3-dimensionally reduced graphene oxide aerogel, and $(NH_4)_2V_6O_{16}\cdot 1.5H_2O$, offered lifetimes of 6000 and 9000 cycles, respectively. The calcium-storage mechanism of $(NH_4)_2V_6O_{16}$ involved a single-phase solid solution reaction. The crystal structure of $(NH_4)_2V_6O_{16}$ and the Ca^{2+} storage site of $(NH_4)_2V_6O_{16}$ were determined.

A layered composite cathode material, comprising $BaV_6O_{16}\cdot 3H_2O$ and graphene oxide, offered a specific capacity of 285.72mAh/g at 50mA/g after 50 cycles, together with a long cycle life[97]. It benefited from a large layer spacing and a robust structure.

In the search for suitable electrode electrodes, first-principles calculations were used[98] to explore the wide chemical landscape of sodium superionic conductor frameworks having the generic formula, $Ca_xM_2(ZO_4)_3$, where M was titanium, vanadium, chromium, manganese, iron, cobalt or nickel and Z was silicon, phosphorus or sulphur. Calculations were made of the average Ca^{2+} intercalation voltage, the thermodynamic stability at 0K of charged and discharged Ca-doped compositions and the migration barriers of metastable Ca-doped compositions. The results showed that $Ca_xV_2(PO_4)_3$, $Ca_xMn_2(SO_4)_3$ and $Ca_xFe_2(SO_4)_3$ are promising cathode materials. All of the silicate Ca-doped compositions were however thermodynamically unstable and therefore unsuitable as cathodes.

This oxide is one of the few host materials which can intercalate calcium ions, and it has long been difficult to identify suitable electrodes for rechargeable Na/Ca-ion batteries offering a good electrochemical performance. Vanadium pentoxides were the best candidates for service as cathodes in such batteries. The concentration-dependent electrochemical characteristics of sodium and calcium ions in α- and δ-V_2O_5 were therefore examined[99] by means of density functional theory calculations, with Hubbard U corrections (table 36). Many low-energy configurations, arising from the various ionic concentrations, were considered in order to determine the stability of α- and δ-V_2O_5 under Na/Ca intercalation. It was predicted that the α-phase was more stable than the δ-phase during sodium and calcium intercalation. The energy barriers for calcium diffusion in α-V_2O_5 at high concentrations (0.975 to 1.825eV) were higher than those in δ-V_2O_5 (0.735 to 1.385eV). This implied that cycling V_2O_5 only in the δ-phase would improve the performance. Lower surface-to-bulk diffusion barriers of 0.498 and 0.846eV were found for sodium and calcium ion-insertion at the (010) surface. This explained the improved electrochemical properties which were observed for nanostructured V_2O_5 as compared with the bulk equivalents. These results provided essential insights into the thermodynamic and electrochemical responses of V_2O_5 to calcium-ion intercalation. Further examination of the surface-to-bulk diffusion of atoms in α- and δ-V_2O_5 showed that the initial barriers for intercalation at the (100) and (010) surfaces were much lower than for diffusion in the bulk. In the case of the α-(010) structure the barrier to calcium atoms, associated with diffusion from the outermost site of the (010) surface to the inside was 0.846eV. The activation barriers to calcium diffusion in α-V_2O_5 were 1.643eV and 1.827eV for low and high concentrations, respectively. It was suggested that the lower diffusion barriers at the surface were controlled by competing factors: broken bonds at the surface led to instability of the surface sites and led to lower activation barriers. Meanwhile a lack of V-O bonds on the cleaved surface led to a higher intercalation energy and hindered movement of the ions into bulk sites. The calcium bonding energy on the α-(010) surface sites was 2.689eV; higher than the 1.243eV for the bulk site. Similar trends were found for calcium intercalation into α-(100) and δ-(010). The lower surface diffusion barrier was expected to favour ion-intercalation into nanostructured V_2O_5. Differences in the calcium mobility at the surfaces and in the bulk played a key role in the intercalation behaviours of bulk crystalline and nano-structured systems. In addition to the diffusion barriers, the electrostatic potential along the diffusion path was also calculated. A strong correlation existed between the peak locations of the electrostatic potential and the calcium activation energy peaks in α- and δ-V_2O_5. High potentials, and wide oscillations of the electrostatic potential, were detected near to the surface, thus strengthening the bonding between calcium and V_2O_5. It was concluded that the large oscillations of electrostatic potential near to the surface were a predominant cause of the high intercalation energy at the surface. The diffusion barriers (0.73 to 1.39eV) for high concentrations of calcium in δ-(010) were lower than those (0.97 to 1.83eV) in α-(010). The calcium in the α a phase was diffused between neighbouring 8-coordinated sites, via a 3-coordinated transition-state. In the δ-phase, the calcium diffused between neighbouring 6-coordinated sites via a 5-coordinated intermediate state and two 3-coordinated transition sites. This led to a smaller coordination change in the δ-phase than in the α-phase. A marked increase occurred in the mobility (about $10^{-25}cm^2/s$) at high calcium concentrations in the δ-phase at room temperature in bulk δ-V_2O_5, as compared with that (about 10^{-33}cm^2/s) in α-V$_2$O$_5$. The smaller coordination changes in the δ-phase gave rise to lower barriers. The calcium migration at low concentrations was calculated to be 0.86eV. The barrier to calcium migration in bulk δ-V$_2$O$_5$ at low concentrations was 0.46eV.

Table 36. Optimized lattice parameters, volume change of the unit cell, average voltage and formation energy (per functional unit) for calcium-intercalated V$_2$O$_5$

Oxide	a(Å)	b(Å)	c(Å)	ε(%)	V(V)	E(meV)
α-V$_2$O$_5$	3.606	4.478	11.433	-	-	0
δ-V$_2$O$_5$	3.643	9.782	10.895	-	-	269
α-Ca$_{0.333}$V$_2$O$_5$	3.582	4.546	11.524	1.646	3.263	644
α-Ca$_{0.667}$V$_2$O$_5$	3.594	4.618	11.498	3.368	3.017	819
α-CaV$_2$O$_5$	3.585	4.755	11.504	6.223	2.248	209

The effect of the presence of water in an electrolyte upon Ca^{2+} insertion and extraction in of α-V$_2$O$_5$ was investigated[100]. Electrochemical studies showed that the overvoltage steadily decreased as the amount of water increased. Co-insertion of H$^+$ into water-containing electrolyte showed that the amount of co-inserted H$^+$ depended upon the amount of water, and there was not much change in overvoltage reduction. The effect of H$^+$ co-insertion upon the electrochemical properties was therefore concluded to be small. Bulk analysis of the electrolyte suggested that hydration to Ca^{2+}, and the dissociation of Ca^{2+} and anions progressed upon increasing the amount of water in the electrolyte. It was deduced that the presence of water in an electrolyte caused a large change in its structure and that this had significant effects regarding overvoltage and electrochemical properties.

Bilayered Mg$_{0.25}$V$_2$O$_5$•H$_2$O was examined[101] as a stable cathode for rechargeable calcium-ion batteries, and an unexpectedly stable structure was found for Ca^{2+} storage. The interlayer spacing underwent only a small change (about 0.09Å) during Ca^{2+} insertion and extraction. This led to very good cycling stability, with 86.9% capacity retention after 500 cycles.

The use of double-sheet vanadium oxide, V$_2$O$_5$•0.63H$_2$O, as a high-performance cathode material for calcium-ion batteries was demonstrated[102]. The oxide was prepared by electrochemical oxidation on a graphite-foil substrate. The material offered a reversible capacity of 204mAh/g at a 0.1C rate in an aqueous electrolyte, with an average discharge voltage of $2.76V_{Ca/Ca2+}$, and a capacity retention of 86% after 350 cycles. The reaction mechanism involved a combination of diffusion-controlled intercalation and surface-limited pseudo-capacitance reactions.

Bilayered $Ca_{0.28}V_2O_5 \cdot H_2O$ cathodes offered a capacity of 142mAh/g at about $3.0V_{Ca/Ca2+}$ together with good cyclability[103]. The material underwent irreversible structural transformation into a 2-fold superstructure during initial charging. The intercalation mechanism was unique in that, upon charging, complete calcium extraction occurred from every 2 interlayers with only a fraction of calcium ions remaining in the other interlayers. During discharging, calcium ions were irregularly inserted into the interlayers and led to stacking faults. The charge-discharge cycle was highly reversible.

Crystalline-water free β-$Ca_{0.14}V_2O_5$ was reported[104] to be a suitable cathode material for calcium-ion batteries. Unlike layered α-V_2O_5 and δ-$Ca_xV_2O_5 \cdot nH_2O$, with their limited capacity, the β-phase offered a reversible capacity of some 247mAh/g. This corresponded to the insertion and extraction of Ca^{2+} between $Ca_{0.14}V_2O_5$ and $Ca_{1.0}V_2O_5$. The initial insertion of Ca^{2+} into $Ca_{0.14}V_2O_5$ caused a small shift in the oxygen atoms which surrounded hepta-coordination sites and created penta-coordinated sites which were then partially filled to give $Ca_{0.33}V_2O_5$. Further insertion occurred via the step-wise occupation of up to 50% of the neighbouring hexa- and tetra-coordination sites to give $Ca_{0.67}V_2O_5$ and $Ca_{1.0}V_2O_5$. Rearrangement of the oxygen atoms in $Ca_{0.14}V_2O_5$ minimized dimensional changes and led to a high cyclic stability during repeated charge and discharge cycles. A remarkable electrochemical behaviour of full cells which contained a $Ca_{0.14}V_2O_5$ cathode and a potassium-metal anode in $Ca^{2+}/K+$ hybrid electrolytes was demonstrated. This was due to the inertness of K^+ insertion into $Ca_{0.14}V_2O_5$ and the absence of calcium plating or stripping. The cyclic stability and capacity of $Ca_{0.14}V_2O_5$ was not impaired in hybrid electrolytes, and this made it a viable cathode for calcium-ion batteries.

A new form of multivalent calcium-ion thermal charging cell was based[105] upon by incorporating the principles of calcium-ion batteries into a thermoelectric system which offered a thermopower of 25.2mV/K via the synchronous thermo-extraction effect of oleic acid-treated $Ca_{0.24}V_2O_5 \cdot H_2O$ electrodes and the thermodiffusion effect of the electrolyte. The calcium-ion energy carrier, favourable aqueous $Ca(CF_3SO_3)_2$ electrolyte, and easy oleic acid modification ensured a possibly higher voltage and rapid diffusion kinetics, and preserved vanadium species from dissolution. The resultant device achieved a thermal voltage of 1.149V and a Carnot-relative efficiency of 24.42% at a temperature-difference of 45K.

An aqueous calcium-ion battery was created[106] which used orthorhombic, trigonal and tetragonal polymorphs of molybdenum vanadium oxide as a host for calcium ions. The orthorhombic and trigonal oxides out-performed the tetragonal structure because large hexagonal and heptagonal tunnels are commonly present in those crystals and provide easy paths for calcium-ion diffusion. In the case of the trigonal oxide, a specific capacity of about 203mAh/g was obtained at 0.2C and a 60mAh/g capacity at 20C. The open tunnels of the trigonal and orthorhombic polymorphs also promoted cyclic stability and reversibility.

Cathodes of V_2O_5 were pre-intercalated with cobalt, manganese or nickel ions and used to explore the coupling effect between guest ions and the host material[107]. Density functional theory simulations showed that the M^{2+} ion interacted with the V-O chains via M^{3d}-O^{2p} covalent bonds. X-ray absorption fine-structure data showed that the Ni-O interatomic distance (1.56Å) was shorter than Co-O (1.60Å) and Mn-O (1.72Å). This suggested that the M-O band-type, with differing degrees of covalency, could optimize the VO_x polyhedron and the local electronic structure. The NiVO cathode material which had the smallest layer-spacing had a higher redox voltage and a better rate/cycling performance for Ca^{2+} storage. This indicated that nickel had a greater tendency to attract electrons and tightly bonded with V-O layers so as to furnish a reliable diffusion-channel for Ca^{2+} ions. Pre-intercalation studies showed how the layer-spacing and the physicochemical properties of the intercalant affects the electrochemical process.

An acetonitrile-water electrolyte was described[108] in which the significant lubricating and shielding effect of the water solvent encouraged the rapid transport of bulky Ca^{2+}. This contributed to a large-capacity storage of Ca^{2+} in layered vanadium oxides of the form, $Ca_{0.25}V_2O_5 \cdot nH_2O$. The acetonitrile component meanwhile suppressed the dissolution of vanadium species during repeated Ca^{2+}-ion uptake and release and imparted a robust cycle life to the cathode. The water molecules were stabilized by mutual hydrogen bonding, with acetonitrile molecules equipping the aqueous hybrid electrolyte with a high electrochemical stability. When using this aqueous hybrid electrolyte, the $Ca_{0.25}V_2O_5 \cdot nH_2O$ electrode offered a specific discharge capacity of 158.2mAh/g at 0.2A/g, a capacity of 104.6mAh/g at a rate of 5A/g and a capacity retention of 95% after 2000 cycles at 1.0A/g. The reversible extraction of Ca^{2+} from the gap of VO polyhedral layers was accompanied by a reversible V-O and V-V skeleton-change and a reversible change in the layer spacing. The ionic diffusion kinetics of the $Ca_{0.25}V_2O_5 \cdot nH_2O$ cathode were calculated (figure 14) with the aid of the galvanostatic intermittent titration

technique during the charging process. The apparent chemical diffusion coefficient of
Ca^{2+} in the material was calculated, for the aqueous hybrid electrolyte, to range from 2.06
x 10^{-11} to 7.96 x 10^{-15}cm^2/s. These values were consistently higher than those for a pure
acetonitrile electrolyte; especially at the end of the charging process. This confirmed the
lubricating and shielding effect of water molecules in accelerating the diffusion of Ca^{2+}
ions.

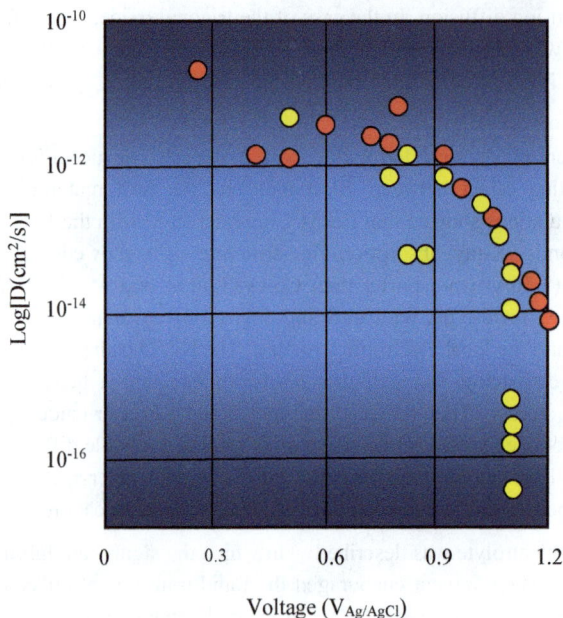

*Figure 14. Calculated diffusion coefficients in Ca$_{0.25}$V$_2$O$_5$•nH$_2$O electrode corresponding
to an aqueous hybrid electrolyte (red) and an acetonitrile electrolyte (yellow)*

Monoclinic $Cu_3(OH)_2V_2O_7 \cdot 2H_2O$, coated with reduced graphene oxide, was prepared by
using a simple one-step surfactant-free co-precipitation method at room temperature and
was studied[109] as a cathode material for calcium-ion batteries. This led to a discharge
capacity of 189.9mAh/g and a rate performance of 55.4mAh at 1000mA/g together with
84.2% capacity retention over 1000 cycles. The Ca^{2+} storage mechanism was based upon
a conversion reaction.

Calcium-ion batteries have a theoretical specific capacity of up to 1337mAh/g, and reversible calcium-ion intercalation was studied in a layered potassium vanadate, $K_2V_6O_{16} \cdot 2.7H_2O$, cathode[110]. The potassium vanadate nano-wires were obtained by using a simple hydrothermal process and had lengths of 0.5 to 1.5μm and widths of 70 to 100nm. They offered an initial capacity of 113.9mAh/g at a current density of 20mA/g, and a capacity retention of 78.30% at 50mA/g after 100 cycles.

Table 37. Ionic conductivity of $Na_2V_6O_{16} \cdot 2H_2O$ electrode in 1M $Ca(ClO_4)_2$ electrolyte and various solvents

Solvent	Conductivity(mS/cm)
H_2O	120
acetonitrile	65.3
propylene carbonate	20.5
dimethyl carbonate	16.7
1,2-dimethoxyethane	14.9
N,N-dimethylformamide	14.1
dimethyl sulfoxide	8.74

Calcium-ion batteries suffer from sluggish kinetics because of the large solvation structure and high de-solvation energy of Ca^{2+} ions. An approach to solvation regulation was proposed[111] which was based upon the donor-number. This enabled the easy-de-solvation and rapid storage of Ca^{2+} in sodium vanadate, $Na_2V_6O_{16} \cdot 2H_2O$. One solvent with a low donor-number was propylene carbonate, and this formed the first solvation shell of calcium ions with a weak binding energy and small shell structure. This then facilitated the migration of Ca^{2+} within the electrolyte (table 37). A low donor-number solvent preferentially de-solvated at the cathode/electrolyte interface, thus promoting the insertion of Ca^{2+} into the $Na_2V_6O_{16} \cdot 2H_2O$ electrode. There was a highly reversible uptake and release of Ca^{2+} in the $Na_2V_6O_{16} \cdot 2H_2O$ cathode, together with a change in the V-O separation of the coordination structure. This cathode offered a capacity of 376mAh/g at 0.3A/g and rate performance of 151mAh/g at 5A/g. The propylene carbonate formed the first solvation shell of calcium ions with weak binding energy and small structure after hybridization with H_2O. Steric effects and ionic conductivity indicated rapid migration of calcium ions in this electrolyte. The redox potential indicated however that Ca^{2+} could be de-solvated and inserted into the $Na_2V_6O_{16}\bullet2H_2O$ electrode at a relatively high potential. The propylene carbonate, because of its weak binding to Ca^{2+}, could promote its preferential de-solvation, with a small amount of H_2O from the solvation shell being inserted into the interlayer so as to expand and stabilize the interlayer; thus ensuring cyclic stability.

An organic molecular intercalation technique was proposed[112] in which V_2O_5, regulated with quinoline, pyridine and water molecules was studied as a cathode material which provided rapid ion-diffusion channels, a large storage capacity and a high conductivity for calcium ions. The V_2O_5-quinoline choice offered the greatest interplanar spacing (1.25nm) and the V-O chains were linked by organic molecules via hydrogen bonding. This stabilized the crystal structure. This cathode offered a specific capacity of 168mAh/g at 1A/g and 80% capacity retention after 500 cycles at 5A/g. X-ray diffraction and absorption spectroscopy revealed a reversible Ca^{2+} order-disorder transformation mechanism which exploited the many active sites which imparted a high capacity, plus rapid reaction kinetics.

A pre-intercalation study was used[113] to improve the cathode performance of polyaniline-V_2O_5. The polyaniline pre-intercalation improved the diffusion kinetics and capacity by increasing the distance (13.8Å) between the V-O layers and markedly increased the material's conductivity. The intercalation of aniline as additional active sites also affected the redox reaction and provided additional capacity. Density functional theory calculations showed that aniline intercalation improved the conductivity and reduced the electrostatic repulsion between Ca^{2+} ions and the V–O layers. Due to the large layer spacing and additional active sites, the material offered a specific capacity of 205mAh/g at 100mA/g.

A superlattice-like poly 3,4-ethylenedioxythiophene-V_2O_5 heterostructure was constructed[114] by using an *in situ* self-assembly method and was used as a cathode. Within the heterostructure, the poly 3,4-ethylenedioxythiophene enlarged the V_2O_5 interlayers and exposed numerous active sites for efficient Ca^{2+} absorption and transport. It also acted as a linkage between neighbouring layers and limited the V_2O_5 volume change during Ca^{2+} insertion and extraction. Due to the synergistic effect of these factors, the electrode had a longer lifespan and greater rate capability than that of V_2O_5. The electrode offered a capacity of 157.2mAh/g at 1A/g and good cycle stability for over 7000 cycles at a current density of 20A/g. It offered a rate capability of 129.0mAh/g, even at 30A/g. Electrochemical studies showed that the insertion and extraction of Ca^{2+}

in this material was highly reversible and was associated with interlayer contraction and expansion.

A sodium-ion pre-intercalated layered $Na_{0.33}V_2O_5$ nano-wire was proposed[115] as a binder-free stable cathode for rechargeable calcium-ion batteries. The pre-intercalated sodium ions acted as supportive structures, linked neighbouring layers, increased the interlayer distance and partially converted pentavalent vanadium cations into tetravalent states. The latter could in turn reduce electrostatic interactions, increase electronic conductivity, boost calcium-ion insertion and extraction and aid structural stability. A hybrid electrolyte which comprised water and tetraethylene glycol dimethyl ether could promote the rapid transport of Ca^{2+} while inhibiting vanadium-species dissolution via its weak effect on solvation. The electrodes offered a specific discharge capacity of 196.3mAh/g at 0.05A/g, a rate capacity of 108.8mAh/g at 3.0A/g and an extended durability of over 2000 cycles.

The electrochemical and kinetic properties of Zn^{2+} and Ca^{2+} ions in aqueous electrolytes were investigated[116] by using a monocrystalline V_2O_5-pyridine material which had a stable and large interlayer spacing. The specific discharge-capacity of 1M $Zn(ClO_4)_2$ aqueous solution was 247.3mAh/g at 0.3A/g. This was much higher than the 158.4mAh/g which was observed for 1M $Ca(ClO_4)_2$. In an aqueous zinc-ion battery, H^+ was first intercalated, followed by the generation of $Zn_4(OH)_7ClO_4$. Finally, H^+ and Zn^{2+} were co-intercalated. In the case of an aqueous calcium-ion battery, H^+ predominated in the intercalation process. Six times more Zn^{2+} than Ca^{2+} was intercalated into the pyridine material, due to its smaller radius and higher intercalation potential: $-0.34V_{Ag/AgCl}$ for Zn^{2+} and $-0.65V_{Ag/AgCl}$. This led to a higher specific capacity. Density functional theory calculations indicated a lower (-6.67eV) intercalation energy for Ca^{2+} than for Zn^{2+} (-1.85eV), and this explained the lower intercalation potential of Ca^{2+}.

Layered $H_2V_3O_8$ with Zn^{2+} pre-insertion was proposed[117] as a high-rate and durable cathode material for calcium-ion batteries. The presence of Zn^{2+} and H_2O pillars could expand the interlayer spacing to 1.8nm, and thus favour diffusion of the large Ca^{2+} ions. The formation of Zn-O bonds facilitated electron transfer and increased electrical conduction. The cathode could therefore offer a capacity of 213.9mAh/g at 0.2A/g and 78.3% retention over a lifespan of 1000 cycles at 5A/g. Density functional theory calculations indicated that the Zn^{2+} moved during Ca^{2+} intercalation, thus lowering the diffusion energy-barrier and facilitating Ca^{2+} diffusion. An aqueous calcium-ion cell was assembled as a proof-of-principle (figure 15). It comprised a 3,4,9,10-perylenetetracarboxylic di-anhydride anode and a $Zn-H_2V_3O_8$ cathode with 1mol/l $Ca(ClO_4)_2$ tetraethylene glycol dimethyl ether:H_2O, in a 4:1 ratio, as the electrolyte.

Given that $Zn-H_2V_3O_8$ provided a better rate capability than did 3,4,9,10-perylenetetracarboxylic di-anhydride at a low current density, a mass ratio of 3,4,9,10-perylenetetracarboxylic di-anhydride to $Zn-H_2V_3O_8$ of 1.2:1 was used. With a view to the voltage-matching of anode and cathode, the operating voltage range of the full cell was set at 0 to 1.7V. A typical cyclic voltammetry curve of the full cell produced 3 consecutive pairs of reduction/oxidation peaks. Galvanostatic charge-discharge tests, performed within this voltage range, yielded specific capacities for $Zn-H_2V_3O_8$ of 129.6, 107.7, 100.6, 93.2, 83.7, 74.2 and 65.7mAh/g at 0.1, 0.15, 0.2, 0.3, 0.5, 0.75 and 1A/g, respectively. These values corresponded to 58.9, 49.0, 45.7, 42.4, 38.0, 33.7 and 29.9mAh/g as based upon the total mass of the 3,4,9,10-perylenetetracarboxylic di-anhydride and $Zn-H_2V_3O_8$. The cyclic performance of the full cell offered a specific capacity of 65.7 at 1A/g for 1000 cycles without significant degradation.

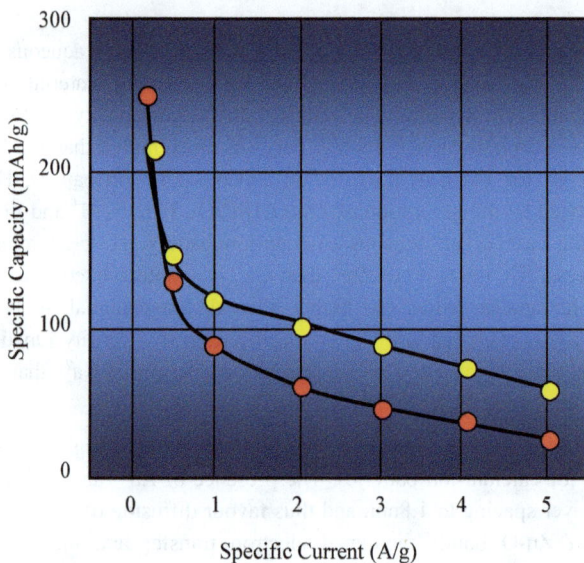

Figure 15. Discharge capacities of $H_2V_3O_8$ (yellow) and $Zn-H_2V_3O_8$ (red)

It was demonstrated[118] that $FeV_3O_9 \cdot 1.2H_2O$ is a good calcium-ion battery cathode material which offers a reversible discharge capacity of 303mAh/g, together with good cycling stability and an average discharge voltage of about $2.6V_{Ca/Ca2+}$. The material was prepared by using a simple co-precipitation method. The bulk intercalation of calcium into the host lattice greatly contributed to the total capacity at lower rates, but became comparable to that due to surface adsorption at higher rates.

Materials having a large interlayer distance offer good specific capacities, and a 1-dimensional structure with adequate calcium-ion passages permit rapid and reversible intercalation. High-yield rapid synthesis of 1-dimensional metal oxides, LiV_3O_8, KV_3O_8, CaV_2O_6 and $CaV_6O_{16} \cdot 7H_2O$ was possible[119] by using a molten-salt method. The 1-dimensional $CaV_6O_{16} \cdot 7H_2O$ offered a capacity of 205mAh/g and a long cycle life with better than 97% of capacity-retention after 200 cycles at 3.0C. There was a figure of 117mAh/g at 12C for calcium-ion intercalation.

Metahewettite layered vanadium oxide, with Ca^{2+} pillars and water lubrication, was studied[120] as a cathode material for calcium-ion batteries. Because of the large interlayer spacing, the pillars and the water-lubrication effect, the $CaV_6O_{16} \cdot 2.8H_2O$ could offer a discharge capacity of 175.2mAh/g at 50C and of 131.7mAh/g at room temperature. It also offered a life of 1000 cycles and a rate performance of up to 1000mA/g in an organic electrolyte. X-ray diffraction and *in situ* Fourier transform infra-red spectroscopic data revealed the occurrence of single-phase Ca^{2+} insertion and extraction. Density functional theory calculations indicated that the Ca^{2+} ions tended to diffuse along the b-direction, with an energy barrier of 0.36eV.

The bronze, $K_{0.5}V_2O_5$, was proposed[121] as a cathode material for calcium-ion batteries. During initial charging of the electrode, the potassium ions were replaced by 0.12mol of calcium. The calcium-exchanged electrode then exhibited a reversible discharge capacity of some 100mAh/g at a 0.1C rate at room temperature, with an average voltage of about $3.0V_{Ca/Ca2+}$. The reversible capacity was higher than that of cathode materials without crystalline water. The latter tends to cause unwanted reactions on the anode.

Most cathode materials suffer from a low capacity or cyclability in 'dry' non-aqueous electrolytes, and materials which offered a capacity greater than 100mAh/g contained crystalline water. The latter was thought to be pivotal with regard to structural stability and the facilitation of calcium diffusion, but β-$Ag_{0.33}V_2O_5$ was proposed[122] as a high-capacity cathode material for calcium-ion batteries, in the absence of crystalline water. Following the initial activation process, this material exhibited a capacity of 179mAh/g at about $2.8V_{Ca2+/Ca}$ in the ninth cycle, together with a reasonable cycling performance. The activation process was attributed to a replacement reaction between the silver and calcium ions. This material thus confirmed that crystalline water is not an essential feature of calcium-ion battery electrode materials offering a high capacity.

Ammonium vanadium oxide, $NH_4V_4O_{10}$, was chosen[123] as an efficient high-capacity cathode for calcium-ion batteries. Conventional hydrothermal processing produced a $NH_4V_4O_{10}$ cathode which exhibited an initial capacity of 125mAh/g at 0.1A/g. The process unfortunately produced sizes which ranged from hundreds of nanometres to a few microns. This limited the electrochemical performance. Uniform rod-like $NH_4V_4O_{10}$ particles which were some 100nm in width could however be produced by adding sodium dodecylbenzenesulfonate surfactant as a soft template during preparation. The surfactant reduced the crystal size and also exerted a Na-doping effect which increased the proportion of V^{4+}/V^{5+} active sites. The Na-doped $NH_4V_4O_{10}$ electrode then offered an initial capacity of 150mAh/g and retained capacity due to coulombic efficiencies of 90 to 95% with no appreciable fading after 100 cycles, in a 3-electrode system. X-ray diffraction revealed the formation of new phases during the migration of calcium ions, and the small change in lattice-plane separation suggested that $NH_4V_4O_{10}$ could offer a stable electrochemical behaviour during prolonged cycling. A full-cell study showed that Na-doped $NH_4V_4O_{10}$ electrodes offered a maximum discharge capacity of 75mAh/g, with a coulombic efficiency of about 80% and a 100% capacity-retention after 100 cycles.

An investigation was made of zirconium-doped $NH_4V_4O_{10}$ for potential use as a high-voltage cathode for rechargeable calcium-ion batteries at room temperature[124]. It exhibited an initial discharge capacity of 78mAh/g, with an average discharge voltage of about $3.0V_{Ca2+/Ca}$ and provided an initial energy-density of 231Wh/kg in a $Ca(CF_3SO_3)_2$-added propylene carbonate electrolyte. The zirconium-doped $NH_4V_4O_{10}$ host offered a high rate-capability and good cycling stability, with 0.021% capacity-decay per cycle over 500 cycles. The capacity retention was some 89%, with a maintained coulombic efficiency of about 100%.

Crystalline-water free $(NH_4)_2V_7O_{16}$ was proposed as a host material[125]. When prepared by using a microwave-assisted hydrothermal method, it had a layered structure with stacked V_7O_{16} layers and interlayer ammonium ions which were hydrogen-bonded to adjacent oxygen atoms. Reversible electrochemical intercalation of Ca^{2+} ions into $(NH_4)_2V_7O_{16}$ led to a reversible capacity of 89mAh/g and an average discharge voltage of about $3.21V_{Ca/Ca2+}$. Although the material had a poor rate capability and cycling performance, it exhibited an unique reaction mechanism. During initial charging, an irreversible structural change occurred which removed all of the ammonium ions and inserted a small number of calcium ions so as to form $Ca_{0.37}V_7O_{16}$. This suggested that an ion-exchange reaction occurred between calcium and ammonium ions. Further cycles involved the reversible co-insertion and co-extraction of calcium and ammonium ions. Without interlayer cations the V_7O_{16} lacked structural stability. This highlighted the importance of co-intercalation of ammonium and carrier ions with regard to reversible cycling.

Metals

There is considerable interest in multivalent cation batteries, such as those based on magnesium, calcium or aluminium. In each case, the choice of the metal anode is a key factor. It had been shown some time before that calcium can be plated and stripped, if only at 75 to 100C, at capacities of the order of $0.165mAh/cm^2$, in the presence of side-reactions. It was demonstrated anew[126] that calcium can be plated and stripped at room temperature with capacities of $1mAh/cm^2$ at a rate of $1mA/cm^2$, with a polarization of about 100mV and for more than 50 cycles. The predominant product is calcium, plus a small amount of CaH_2 that forms by reaction between the deposited calcium and the electrolyte, $Ca(BH_4)_2$ in tetrahydrofuran. This replaced the usual reactions which occur in most electrolyte solutions to form $CaCO_3$, $Ca(OH)_2$ and calcium alkoxides. The CaH_2 protects the calcium metal under open-circuit conditions. This demonstrated that appreciable amounts of calcium can be plated and stripped at room temperature, with low polarization.

beryllium

First-principles calculations showed[127] that the highly conductive planar hexacoordinated Be_2Al and Be_2Ga monolayers could serve as anodes for alkali/alkaline metal-ion batteries. Due to a favourable adsorption-energy landscape, the barriers to diffusion of lithium, sodium, potassium and calcium atoms on the Be_2Al and Be_2Ga surfaces were 0.116eV, 0.075eV, 0.046eV and 0.270eV, and 0.044eV, 0.025eV, 0.014eV and 0.217eV, respectively. These values promised high fast rate capabilities. The Be_2Al and Be_2Ga monolayers exhibited mean open-circuit voltages of 0.049V and 0.104V for Li^+, 0.234V and 0.278V for Na^+, 0.249V and 0.227V for K^+, and 0.120V and 0.124V for Ca^{2+}, respectively. Due to stable double-sided multilayer adsorption, the Be_2Al and Be_2Ga monolayers could offer reversible specific capacities of 2382 and 1222, 4764 and 1833, 2382 and 1222, and 7146 and 3665mAh/g for lithium, sodium, potassium and calcium atoms, respectively.

calcium

It is difficult to plate and strip, efficiently and reversibly, calcium-metal anodes in organic electrolytes. The inorganic components of the solid/electrolyte interphase which forms via the decomposition of the electrolyte tend not to allow for the diffusion of calcium ions. Density functional theory and *ab initio* molecular dynamics simulations were used[128] to show that the use of a pre-formed artificial solid/electrolyte amorphous interphase layer could prevent electrolyte decomposition. It was first shown that calcium is able to intercalate into an amorphous layer up to $Ca_{1.5}Al_2O_3$ (figure 16) and diffuse through within a reasonable time-scale. Calculation of the density-of-states showed that the system remained insulating up to the equilibrium stoichiometry. Molecular dynamics simulations for a realistic organic electrolyte showed that a layer could completely prevent the decomposition of solvent molecules. This was expected to provide a route to the creation of efficient rechargeable calcium-ion batteries.

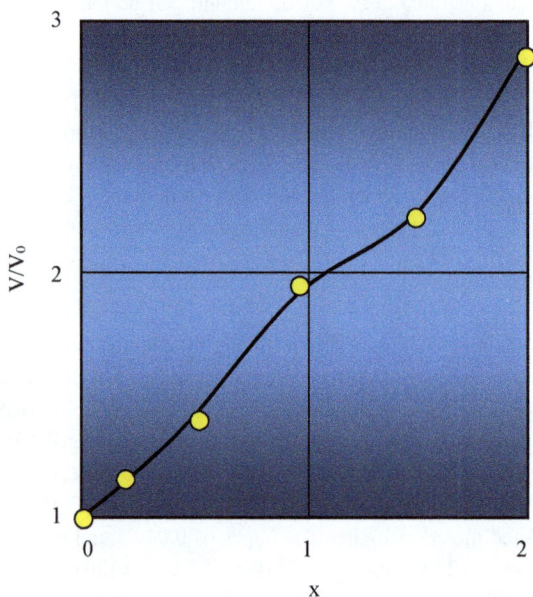

Figure 16. Volume expansion of $Ca_xAl_2O_3$, relative to the volume of amorphous $Al_{40}O_{60}$

tin

The feasibility of using calcium-tin alloy anodes for calcium-ion batteries was demonstrated[129]. Crystallographic and microstructural studies revealed that tin which formed via electrochemical de-alloying of the calcium-tin alloy exhibited unique properties, and the tin which was formed later underwent reversible calcination or de-calciation as $CaSn_3$. Coin-cells were created which had an organic (1,4-polyanthraquinone) cathode in an electrolyte which consisted of 0.25M calcium tetrakis(hexafluoro-isopropyloxy)borate in dimethoxyethane. These cells were charged and discharged for 5000 cycles at 260mA/g and maintained a capacity of 78mAh/g with respect to the organic cathode. A high-voltage (4.45V) calcium-ion battery cell which used tin as the anode exhibited a cyclability of more than 300 cycles.

Table 38. Candidate high-performance calcium-alloy anodes with relaxed voltage constraint ($V_{threshold} = 0.1V$): V_{Ca}: calcination voltage, C_g: gravimetric capacity, C_v: volumetric capacity, E_d: energy density, E_s: specific energy, ε: volume expansion

Addition	V_{Ca}(V)	C_g(mAh/g)	C_v(mAl/ml)	E_d(Wh/l)	E_s(Wh/kg)	ε(%)
Ag	0.24	828	8866	42209	3943	383.18
Al	0.28	1845	4910	23167	8702	201.22
As	0.75	1431	8271	35176	6085	287.48
Au	0.33	816	15830	73911	3811	683.38
Cd	0.16	1119	10130	49030	5414	422.89
Cd	0.26	715	6227	29542	3393	267.43
Cu	0.16	843	7581	36676	4080	315.98
Ga	0.22	1957	11680	55842	9355	474.07
Ge	0.47	1476	7864	35622	6685	274.47
In	0.21	1400	10381	49682	6702	372.00
Pb	0.28	776	8897	41965	3660	280.85
Pd	0.30	1511	18141	85225	7098	769.50
Pt	0.50	687	14738	66261	3088	573.37
Sb	0.67	880	5778	25000	3809	192.56
Si	0.35	3817	8892	41310	17732	318.64
Sn	0.53	903	5249	23465	4036	183.64
Tl	0.15	787	9185	44536	3815	316.01
Zn	0.16	1366	9816	47472	6607	438.78

The exact reactions proposed were: $Ag+(5/3)Ca \rightarrow (1/3)Ca_5Ag_3$, $Al+(13/14)Ca \rightarrow (1/14)Ca_{13}Al$, $As+2Ca \rightarrow Ca_2As$, $Au+3Ca \rightarrow Ca_3Au$, $(1/3)CaHg_3+(8/3)Ca \rightarrow Ca_3Hg$, $Cd+(3/2)Ca \rightarrow (1/2)Ca_3Cd_2$, $Cu+Ca \rightarrow CaCu$, $Ga+(28/11)Ca \rightarrow (1/11)Ca_{28}Ga_{11}$, $Ge+2Ca \rightarrow Ca_2Ge$, $In+3Ca \rightarrow Ca_3In$, $Pb+3Ca \rightarrow Ca_3Pb$, $Pd+3Ca \rightarrow Ca_3Pd$, $Pt+(5/2)Ca \rightarrow (1/2)Ca_5Pt_2$, $Sb+2Ca \rightarrow Ca_2Sb$, $Si+2Ca \rightarrow Ca_2Si$, $Sn+2Ca \rightarrow Ca_2Sn$, $Tl+3Ca \rightarrow Ca_3Tl$, $Zn+(5/3)Ca \rightarrow (1/3)Ca_5Zn_3$

Table 39. Candidate high-performance calcium-alloy anodes with relaxed voltage constraint ($V_{threshold}$ = 0.53V): V_{Ca}: calcination voltage, C_g: gravimetric capacity, C_v: volumetric capacity, E_d: energy density, E_s: specific energy, ε: volume expansion

Addition	V_{Ca}(V)	C_g(mAh/g)	C_v(mAl/ml)	E_d(Wh/l)	E_s(Wh/kg)	ε(%)
As	0.75	1431	8271	35176	6085	287.48
Au	0.56	454	8795	39066	2014	340.43
Bi	0.59	513	5031	22200	2264	160.35
Ge	0.67	738	3932	17013	3193	118.83
Pd	0.54	755	9070	40445	3368	341.44
Pt	0.67	458	9825	42533	1982	324.58
Sb	0.67	880	5778	25000	3809	192.56
Si	0.55	1908	4446	19784	8492	137.85
Sn	0.59	527	3062	13496	2321	94.94

The exact reactions proposed were As+2Ca→Ca$_2$As, Au+(5/3)Ca→(1/3)Ca$_5$Au$_3$, Bi+2Ca→Ca$_2$Bi, Ge+Ca→CaGe, Pd+(3/2)Ca→(1/2)Ca$_3$Pd$_2$, Pt+(5/3)Ca→(1/3)Ca$_5$Pt$_3$, Sb+2Ca→Ca$_2$Sb, Si+Ca→CaSi, Sn+(7/6)Ca→(1/6)Ca$_7$Sn$_6$

Calciation of the tin was observed to cease at Ca$_7$Sn$_6$, a surprisingly lower composition than that of compounds such as Ca$_2$Sn. The electrochemistry of the tin-calciation reaction was investigated[130] using density functional theory. This identified the threshold voltages which governed the limits of calciation. That information was then used to design a 4-step screening process and density functional theory was used to search for anode materials possessing superior properties. Metalloids (silicon, antimony, germanium) and post-transition metals (aluminium, lead, copper, cadmium, CdCu$_2$) were predicted to be promising candidates for anodes (tables 38 and 39).

Calcium-ion batteries also suffer from poor reversibility and a short lifespan of calcium-metal anodes. Solvation manipulation can improve the plating/stripping reversibility of calcium-metal anodes by increasing the de-solvation kinetics of calcium ions in the electrolyte. The introduction of lithium salts can considerably change the electrolyte structure by reducing the coordination-number of calcium ions in the first solvation shell. A coulombic efficiency of up to 99.1% was achieved[131] for the galvanostatic plating/stripping of a calcium-metal anode, together with very stable long-term cycling over 200 cycles at room temperature.

First-principles calculations were used[132] to study the intrinsic and reductive stability of 12 calcium salts with fluorinated aluminate and borate anions. An analysis was made of the decomposition products which were formed on the metal anode surface, because these are important with regard to early-stage solid-electrolyte interphase formation. Anions with appreciable steric hindrance and a high degree of fluorination were intrinsically less stable and were concluded to be unsuitable for calcium salts. Aluminate salts were generally less reactive with calcium anodes than were their borate counterparts; a high degree of fluorination led to lower reductive stability. Calcium fluoride was the most notable decomposition product on the anode surface, while carbide-like products also arose from decomposition of the salts.

A study was made[133] of the stability of 4 calcium salts having weakly coordinated anions, and of 3 solvents which are used in batteries in order to identify candidates for calcium-ion batteries. Born-Oppenheimer molecular dynamics simulations of salt/calcium and solvent/calcium interfaces showed that a tetraglyme solvent and a carborane salt are promising candidates for calcium-ion batteries. Because of the highly reducing nature of the calcium surface, other salts and solvents easily decompose. The microscopic mechanisms of salt/solvent decomposition on the calcium surface were explained by using the time-dependent projected density-of-states, the time-dependent charge-transfer plots and nudged elastic band calculations. This was a useful mechanistic assessment of the dynamic stability of candidate salts and solvents on a calcium surface.

Candidate Electrolytes

Non-lithium|sulphur batteries are promising systems, but there still lacks an understanding of quantitative cell parameters and the mechanisms of sulphur electrocatalytic conversion, which limits progress in developing these batteries. A detailed analysis was made[134] of electrode criteria such as the sulphur mass loading, sulphur content, electrolyte/sulphur ratio and negative/positive electrode-capacity ratio with regard to the specific energy, Wh/kg, of post-lithium metal|sulphur batteries. A critical evaluation was made of progress in investigating electrochemical sulphur conversion via homogeneous and heterogeneous electrocatalytic approaches to non-aqueous Na/K/Mg/Ca/Al|S and aqueous Zn|S batteries.

Ab initio molecular dynamics, non-equilibrium Green's function techniques and density functional theory were used[135] to investigate the electrolyte solvents, tetrahydrofuran and ethylene carbonate, with regard to charge transport as the solvent molecules interacted with an anode surface. The tetrahydrofuran maintained a consistent conductance, although there were some jumps in the conductance which were attributed to molecular rearrangement. The ethylene carbonate underwent a large degree of molecular decomposition, with a corresponding decrease in the conductance of some orders-of-magnitude. Molecular decomposition and early solid-electrolyte interphase formation played a large part in charge transport change as a function of time and temperature.

Cross-linked poly(tetrahydrofuran) networks are useful as organic solvent/oil sorbents which exhibit a high degree of swelling, and as mechanically strong solid polymer electrolyte membranes for solid-state calcium-ion batteries. The control of intra-molecular and intermolecular interactions is important to the modulation of the properties of dynamic networks which involve non-covalent and dynamic covalent bonds. Differences in the dynamics of different polymeric materials, although prepared from the same precursor polymer, remain puzzling. A study was made[136] of the dissimilar dynamic properties of poly(tetrahydrofuran)-based materials which were prepared via intramolecular or intermolecular chemistry to give films of poly(tetrahydrofuran)-based single-chain nanoparticles or cross-linked networks. There were marked similarities in the segmental relaxation time and the glass transition temperature, but clear differences in the dynamic heterogeneities. The differences were attributed to the peculiar morphology of intra-chain collapsed polymer chains.

Table 40. Activation energy of polytetrahydrofuran-epoxy networks

O:Ca	$T_g(K)$	E(eV)
13.8	200	0.229
8.4	201	0.247
4.2	200	0.244
3.8	200	0.219
2.3	201	0.295
1.9	217	0.310
1.1	210	0.328

T_g: glass transition temperature

Calcium is not only a very abundant element, it also offers more than twice as great a volumetric capacity as do monovalent lithium-ion batteries. In comparison with other multivalent ions, such as magnesium, calcium did not become really attractive before the appearance of $Ca(BH_4)_2$-in-tetrahydrofuran as an electrolyte within which calcium could be plated and stripped at room temperature, with little polarization. Because solvation and de-solvation on the cathode or anode were simultaneous, an understanding of the dynamic Ca^{2+} solvation process at the electrode/electrolyte interface is essential when trying to design an electrolyte which is compatible with both electrodes. A review[137] of Ca^{2+} salvation, based upon computational and experimental results, suggested a multi-ion strategy for improving battery performance.

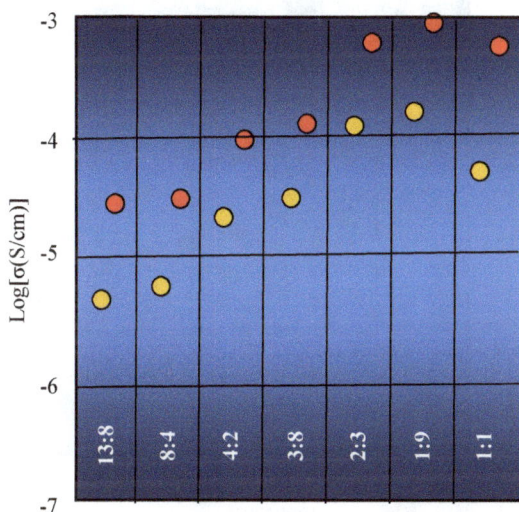

Figure 17. Ionic conductivity of polytetrahydrofuran-epoxy solid polymer electrolytes as a function of temperature. Orange: O:Ca = 1:9, red: O:Ca = 4:2, green: O:Ca = 13:8

Solid networks were produced[138] from polytetrahydrofuran and 3,4-epoxycyclohexylmethyl-3′,4′-epoxycyclohexane carboxylate by means of visible-light initiation of photo cross-linking. The networks were loaded with various amounts of calcium nitrate so as to create solid-polymer electrolytes. The ionic conductivity (table

40, figures 17 and 18) was determined by means of impedance spectroscopy, while the thermal properties were determined by means of thermogravimetric analysis and differential scanning calorimetry. All of the materials were rubber-like and were stable at 30 to 120C. With increasing salt loading, the ionic conductivity of the electrolytes first increased and then decreased. Material having a molar O:Ca ratio of 1.9 offered the highest conductivity (1.14 x 10⁻⁴S/cm) at room temperature and a transference number of 0.359 at 70C. This co-polymer system was the basis of an approach to designing calcium-ion solid electrolytes for solid-state calcium-ion batteries.

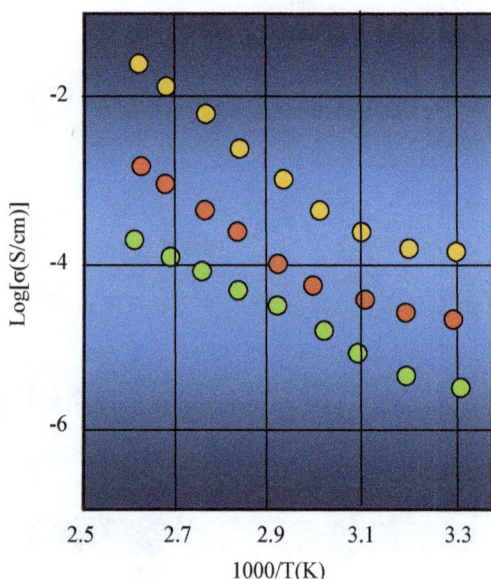

Figure 18. Ionic conductivity of polytetrahydrofuran-epoxy solid polymer electrolytes at 40C (red) and 70C (orange) as a function of the O:Ca ratio

It is possible for calcium to deposit reversibly in the presence of liquid electrolytes. The metal is incompatible with conventional non-aqueous electrolytes and few calcium salts are easily available: calcium tetrafluoroborate, calcium bis(trifluoromethanesulfonyl) imide, calcium nitrate and calcium borohydride. Some, such as calcium perchlorate, are clearly unsafe[139] and would revive the dangers inherent in the use of lithium. Solid polymer electrolytes are generally not suitable because Ca^{2+} diffusion is too slow to be practical, although CaF_2 has been investigated as a solid electrolyte for Ca-Mg, Ca-Bi and Ca-Sn alloys. This fluoride becomes a mixed conductor above 873C. Metal borohydrides of the form, $A_xB_nH_n$, where A was lithium or sodium, x was 1 or 2 and n was 11 or 12 or borohydrides of the form, MB_nH_n, where M was calcium or magnesium and n was 11 or 12, have been proposed as candidates for solid-state batteries. Such metal borohydrides are however unstable and tend to become oxidized at high temperatures. The decomposition products form a protective layer at the electrode interface and offer an improved electrochemical stability. Poly(ethylene-glycol) diacrylate has been used as a polymer electrolyte for calcium batteries. The poly(ethylene-glycol) diacrylate was cross-linked under blue light in the presence of calcium nitrate, and formed a stable network. In general, an increase in the salt content leads to an increase in ionic conductivity. Its decomposition temperature is greater than 130C. PTHF-Epoxy has been used as a polymer electrolyte for calcium batteries, and was prepared via the co-polymerisation of PTHF with a cyclo-aliphatic epoxy that was loaded with Ca^{2+} in the form of calcium nitrate. The properties of gel-polymer electrolytes depend strongly upon the salt content, due to extra solvated Ca^{2+}, and can be stable over a wide temperature range. Various calcium salts have been incorporated into poly(oxyethylene) and used as a solvent-free polymer electrolyte for calcium batteries. The doping of calcium with aluminium, bismuth or rare-earth elements creates calcium vacancies and improves the diffusion kinetics. The calcium oxide passivation layer which usually forms on a calcium anode is highly stable.

An ionic-liquid polymer-gel membrane was proposed[140] for use as both the electrolyte and separator in a calcium-ion battery operating at room temperature. The membrane was prepared via the single-step photo cross-linking of poly(ethylene-glycol) diacrylate in the presence of a calcium salt dissolved in an ionic liquid. This led to room-temperature ionic conductivities of between 10^{-4} and 10^{-3}S/cm, a voltage of about $4V_{Ca/Ca2+}$, a cationic transference number of 0.17, good thermal stability at up to some 300C and full dissociation of the calcium salts in the ionic liquid. A prototype battery exhibited intercalation-based room-temperature operation, with an initial discharge capacity of 140mAh/g.

Although calcium-ion batteries have emerged as a leading contender for sustainable energy storage, there remains a need to develop electrolytes which improve their performance. A gel-polymer electrolyte was based[141] upon polyvinylidene fluoride for calcium-ion conduction. This electrolyte was prepared by combining a polyvinylidene fluoride polymer host, a $Ca(TFSI)_2$ salt and N-methyl-2-pyrrolidone solvent. Fourier-transform infra-red spectroscopy revealed the effect of salt concentration and drying temperature on the degree of salt dissociation in the electrolyte. The concentration of free cations in the electrolyte was mainly related to N-methyl-2-pyrrolidone and as well as polyvinylidene fluoride. This generated a suitable network for ion transport: a liquid electrolyte enclosed within a polymer matrix. Processing conditions such as the drying temperature, which varied the solvent content, played an important role in developing polymer electrolytes that offered optimum electrochemical behaviour. The gel-polymer electrolytes were semicrystalline and were stable at up to 120C; a critical factor for their use in electric vehicles and renewable energy storage systems. The ionic conductivity of the gel-polymer electrolytes (figure 19) exhibited Arrhenius-type behaviour and the total ionic conductivity at room temperature was 0.85 x 10^{-4}S/cm for 0.5M and 3.56 x 10^{-4}S/cm for 1.0M concentrations.

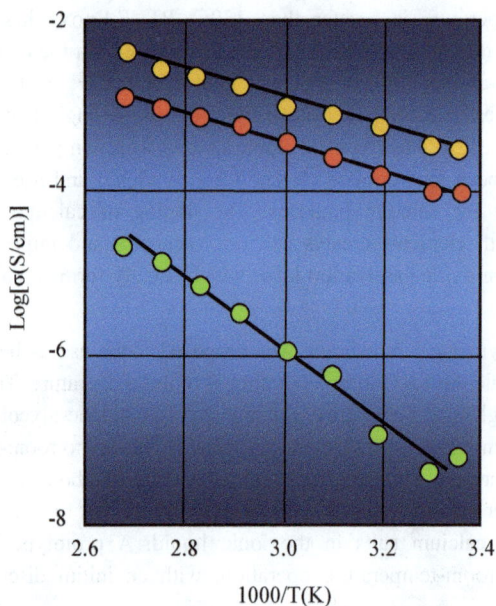

Figure 19. Conductivity at 75C of an electrolyte comprising a polyvinylidene fluoride polymer host, a Ca(TFSI)$_2$ salt and an N-methyl-2-pyrrolidone solvent. Orange: 0.1M, red: 0.5M, green: 1.0M

Candidate cathode materials include transition-metal compounds, which undergo easy dissolution in aqueous solutions but a low conductivity. This hinders their use, but advantage can be taken of the so-called conveyor-effect of conjugated polyaniline in order to produce an oxygen-defective tungstate-linked polyaniline material having a chrysanthemum-like microstructure. Because of a high electronic conductivity that is due to a conductive conjugated polyaniline skeleton, an unbalanced charge distribution due to the defective structure and a reversibly rapid ion intercalation that exploits the open framework and porous chrysanthemum-like microstructure, it offers a high rate capability with a maximum specific capacity of 162.2mAh/g and good cycle stability. It also offers a reversible capacity of 140.4mAh/g and a marked ability to store Ca^{2+}. Assembled calcium-ion batteries therefore exhibit good capacities, energy densities and flexibilities[142]. The Ca^{2+} storage involves interaction via ionic bonds.

Table 41. Conductivity activation energy of poly(ethylene glycol) diacrylate-calcium solid polymer electrolytes

O:Ca	$T_g(K)$	E(eV)
52	220.15	0.225
30	229.95	0.244
20	241.55	0.233
10	276.1	0.282
5	300.15	0.250

T_g: glass transition temperature

In further work, polymer gel electrolytes for calcium-ion conduction were prepared[143] via the photo cross-linking of poly(ethylene-glycol) diacrylate in the presence of solutions of calcium salts in a mixture of ethylene carbonate and propylene carbonate solvents. This led to room-temperature conductivities of between 10^{-5} and 10^{-4}S/cm and an electrochemical stability window of 3.8V, together with full dissociation of the salt and minimal coordination with the poly(ethylene-glycol) diacrylate backbone. Cycling in symmetrical calcium metal cells occurred with increasing overpotentials that were attributed to an interfacial impedance between the electrolyte and the calcium surface which inhibited charge transfer. Calcium could still be plated and stripped to yield high-

Materials Research Forum LLC
https://doi.org/10.21741/9781644903490

purity deposits with no sign of appreciable electrolyte breakdown. This indicated that high over-potentials were associated with an electrically insulating and ion-permeable solid/electrolyte interface.

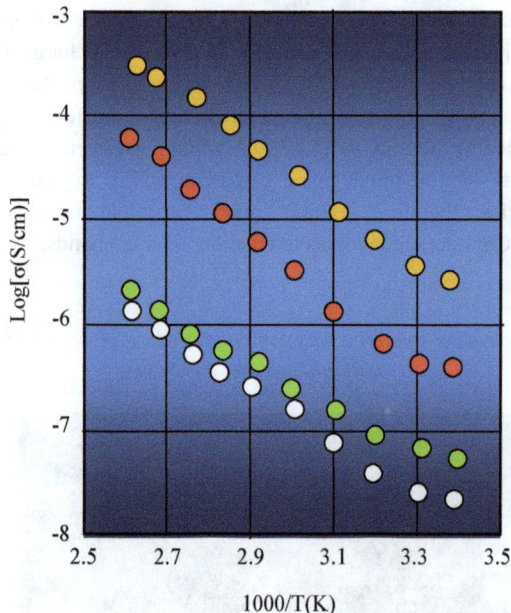

Figure 20. Ionic conductivity of poly(ethylene glycol) diacrylate-calcium solid polymer electrolyte networks. Orange: OE:Ca = 5, red: OE:Ca = 10, green: OE:Ca = pure poly(ethylene glycol) diacrylate, white: OE:Ca = 52

Conventional electrolytes involving oxygen-based coordination chemistries have to be modified in order to ensure more rapid cation transport. An imidazole-based polymer electrolyte having the then-highest known conductivity was described[144]. The polymerization of vinylimidazole in the presence of calcium bis(trifluoromethanesulfonyl)imide salt created a gel electrolyte consisting of a polyvinyl imidazole host which was infused with vinylimidazole liquid. The calcium ions coordinated with imidazole groups and the electrolytes exhibited room-temperature conductivities greater than 1mS/cm. Reversible redox activity in symmetrical calcium cells was detected at 2V over-potentials, together with stable cycles at $0.1mA/cm^2$ and areal capacities of 0.1mAh/cm². The satisfactory properties were attributed to softer coordination, polarizability and a closer coordination of the site distances of the imidazole groups.

A solid polymer electrolyte for calcium-ion conduction was produced[145] by photo cross-linking poly(ethylene glycol) diacrylate in blue light in the presence of a calcium salt so as to form a stable network which offered high thermal stability and an ionic conductivity of 3.4×10^{-4} S/cm at 110C and of 3.0×10^{-6} S/cm at room temperature. The electrolytes remained stable up to about 140C. There was no salt precipitation within the polymer matrix and Raman analysis indicated complexation of the calcium ions with the poly(ethylene glycol) diacrylate network. The ionic conductivities of the poly(ethylene glycol) diacrylate-calcium electrolytes as a function of temperature (figure 20) showed that an increase in salt loading increased the conductivity apart from the composition, EO/Ca = 52. This had a lower conductivity (table 41) than did pure poly(ethylene glycol) diacrylate. This anomaly was attributed to the cross-linking effect of calcium ions in the polymer chain, which impaired the segmental mobility of the samples. When the concentration of calcium nitrate was sufficiently high, the availability of carrier-ions increased and the conductivity therefore increased. The data suggested that the low concentration of calcium nitrate in the sample, EO/Ca = 52, was insufficient to overcome the loss of mobility, with enough ions present to increase the conductivity.

A super-concentrated aqueous electrolyte was initially used[146] in a calcium-ion battery in order to improve the electrochemical performance by decreasing the hydration number and the radius of the calcium ions. Their charge/discharge capacity, with the super-concentrated electrolyte, was about 13% higher than that with a dilute electrolyte. The cycling performance markedly improved in the case of the super-concentrated electrolyte. This was associated with suppression of the structural collapse of the copper hexacyanoferrate which was used as an electrode. The narrow (1.23V) electrochemical stability window of an aqueous electrolyte limits the creation of high energy-density and long cycle-lifetime batteries. A calcium-based water-in-salt aqueous electrolyte was found[147] to overcome the narrow stability-window problem by offering a 2.12V-wide window. This was done by suppressing hydrogen evolution at the anode, and minimizing overall water activity at the cathode. A theoretical analysis indicated that the preferential reduction of salt aggregates led to the formation of a passivation layer at the electrode/electrolyte interface and widened the electrolyte stability window. Raman spectroscopy results indicated that the calcium-ion coordination number, the number of nitrate ions surrounding the calcium ions in an aqueous electrolyte, increases with increasing electrolyte concentration. This then led to a gradual decrease in the hydration number of the calcium ions. A full cell which was based upon this water-in-salt electrolyte offered a good rate capability and a cycling stability with a capacity-loss of just 0.01/cycle, and 80% capacity-retention over 1800 cycles with some 99.99% coulombic efficiency. The full cell provided an energy-density of 232Wh/kg at a power-density of 69W/kg and a current rate of 0.15A/g. At a current rate of 5A/g it offered an energy-density of 182Wh/kg.

Table 42. Cyclic performance of calcium-ion batteries with various electrolytes

Cathode	Anode	Electrolyte	Voltage(V)	C_m(mAh/g)	C_r(%)
$Fe_2(CN)_6$	$NiNH_2$	A	0 to 2	60 at 0.1A/g	92, a
$KNiFe(CN)_6$	ac	B	0.2 to 0.9	35 at 0.25A/g	53, b
CuHCF	ac	C	0 to 1	70 at 0.017A/g	88.6, c
CuHCF	polyimide	D	0.5 to 1.9	25 at 0.4A/g	95, a
ZnHCF	Sn-doped In_2O_3	E	0 to 1	50.9 at 0.5A/g	70, d
$CaCo_2O_4$	V_2O_5	F	-1.2 to 2.5	100 at 40μA	75, e
$NH_4V_4O_{10}$	$MnNH_2$	G	0.01 to 2	75 at 0.1A/g	100, f
$Na_{0.5}VPO_{4.8}F_{0.7}$	ac	H	-1 to 1.5	75 at 0.05A/g	90, g

A: hybrid 3.1M $Ca(ClO_4)_2/(H_2O)_4(AN)_{4.8}$, B: aqueous 1M $Ca(NO_3)_2$, C: aqueous 8.4M $Ca(NO_3)_2$, D: aqueous 2.5M $Ca(NO_3)_2$, E: aqueous 1M $Ca(Cl)_2$, F: organic 1M $Ca(ClO_4)_2$ in AN, G: organic 1M $Ca(ClO_4)_2$ in AN, H: organic 1M $Ca(PF_6)_2$ in EC/PC, AN: acetonitrile, ac: activated carbon, a: 1000 cycles, b: 2000 cycles, c: 5000 cycles, d: 250 cycles, e: 30 cycles, f: 100 cycles, g: 500 cycles, C_m: maximum capacity, C_r: residual capacity.

Highly concentrated water-in-salt electrolytes offer an increased stability by limiting the number of free water-molecules which are present in the system. The electrolytes (table 42) still suffer from a low conductivity and a high viscosity, due to the high concentration of the calcium salt. Hybrid aqueous/organic calcium electrolytes have been developed[148] in order to improve the electrochemical stability window and the long-term reversibility, but with a low electrolyte concentration. The best hybrid-electrolyte comprised calcium perchlorate, water and acetonitrile in a molar ratio of 1.0:4.0:4.8. This led to an electrochemical stability window of about 3V, and the easy insertion of calcium ions into Prussian Green host materials. This offered 90% capacity retention following 1000 cycles. Nickel foam was used as a cathode current-collector in order to determine the feasibility of drop-coated 3-dimensional electrodes. Because of its high porosity the nickel foam substrate could offer a 4 to 8 times higher mass-loading than could traditional foil-type collectors, and led to cells having much higher energy-densities.

Miscellaneous Battery Data

Multivalent ion batteries which are based upon zinc-, calcium-, aluminium- or magnesium-ions are promising due their abundance. Their multivalent chemistry promises the creation of high energy-density storage systems. There remain the problems of low electrochemical reversibility and slow reaction kinetics. There is also a danger of side-reactions with flammable and toxic organic liquid electrolyte, and dendrite growth. Quasi-solid and solid-polymer electrolytes have therefore attracted interest as safer alternatives[149]. Unfortunately only a few calcium-ion electrolyte systems are available which enable reversible plating at room temperature. These include aluminates and borates such as $Ca[TPFA]_2$ and $Ca[B(hfip)4]_2$ where $[TPFA]^- = [Al(OC(CF_3)_3)_4]^-$ and $[B(hfip)4]_2^- = [B(OCH(CF_3)_2)_4]^-$. A common feature of the structure of these compounds is a weakly coordinated anion consisting of a tetra-coordinated aluminium/boron centre and fluorinated alkoxide. There is a dependence of Coulombic efficiency upon their natural tendency to cation-anion coordination. An innovation[150] has been the 1-pot synthesis of new calcium-ion electrolyte compounds: $Ca[Al(tftb)4]_2$, $Ca[Al(hftb)4]_2$, $Ca[Al(hfip)4]_2$ and $Ca[TPFA]_2$ where hfip = $-OCH(CF_3)_2$, tftb = $-OC(CF_3)(Me)_2$, hftb = $-OC(CF_3)_2(Me)$ and $[TPFA]^- = \{Al[OC(CF_3)_3]_4\}^-$.

The diffusion of cations in organic solvent solutions is important to the operation of metal-ion batteries. Pulsed field gradient nuclear magnetic resonance and atomistic molecular dynamic simulations have been used to study the temperature-dependent diffusion of various common liquid electrolytes such as 1M propylene carbonate solutions of metal salts with bis(trifluoromethylsulfonyl)imide or hexafluorophosphate anions. The results[151] revealed an Arrhenius-type temperature dependence of the diffusion coefficients of the propylene carbonate solvent and the anions. The propylene carbonate molecules were the faster species. In the case of monovalent cations such as Li^+, Na^+ and K^+, the propylene carbonate solvent diffusion increased with increasing cation size increased. In the case of divalent cations such as Mg^{2+}, Ca^{2+} Sr^{2+} and Ba^{2+}, the diffusion coefficients decreased as the cation size increased. The anion diffusion in Li-bis(trifluoromethylsulfonyl)imide and Na-bis(trifluoromethylsulfonyl)imide solutions was similar while, in electrolytes with divalent cations, there was a decrease in anion diffusion with increasing cation size. That non-polarizable charge-scaled force fields fitted very well with the experimental values of the anion and propylene carbonate

Materials Research Forum LLC

https://doi.org/10.21741/9781644903490

solvent diffusion coefficients in salt solutions of both monovalent and divalent cations over a range of temperatures. According to calculations of the radial distribution functions between cations, anions and solvent molecules, the increase in propylene carbonate diffusion coefficient with increasing cation size for monovalent cations was due to the large hydration shell of small Li^+ cations because of their strong interaction with the propylene carbonate solvent. In solutions with larger monovalent cations, such as Na^+, and with a smaller solvation shell of propylene carbonate, the diffusion of the latter was faster because of more liberated solvent molecules. In salt solutions with divalent cations, both the anion and propylene carbonate diffusion coefficients decreased as the cation size increased due to an increased cation-anion coordination that was associated with an increase in the amount of propylene carbonate in the cation solvation shell because of the presence of anions.

The electrolytes in calcium-ion batteries exhibit a lack of stability and a degradation which is caused by reduction from the anode. A solid/electrolyte interphase which forms on anodes during operation impedes the flow of electrons from the anode to the electrolyte[152]. A common inorganic compound which is found in the solid/electrolyte interphase is CaF_2 and arises from electrolyte salts such as $Ca(PF_6)_2$. The CaF_2 can exist as crystalline, polycrystalline or amorphous phases in the solid/electrolyte interphase, and differing phases of the same compound can have very different electronic conductivities. Amorphous-phase systems promoted electron-tunnelling in thin CaF_2 films by 1 to 2 orders-of-magnitude, as compared with crystalline and polycrystalline CaF_2 systems. The transport through amorphous structures was such that, in spite of their random structures, their conductances were similar. Analysis of the decay-constant and the low-bias conductance showed that crystalline and polycrystalline CaF_2 afforded a greater protection to the electrolyte than did amorphous CaF_2.

There is an interest in the design of electrolytes for multivalent-ion batteries by tuning the molecular-level interactions of solvate species in the electrolytes. An attempt was made[153] to control calcium-ion speciation in ionic liquid-based electrolytes by designing alkoxy-functionalized cations. It was found that the alkoxy-functionalized ammonium cation, N_{07}^+, bearing 7 ether oxygen atoms could displace the bis(trifluoromethanesulfonyl)imide (TFSI) anion from the calcium-ion coordination sphere, thus facilitating reversible calcium deposition and stripping. Analysis of calcium deposits, and density functional theory calculations, indicated the formation of an organic-rich, inorganic-poor solid electrolyte interphase layer. This aided calcium-ion diffusion rather than passivating the calcium-metal electrode. A prototype Ca/V_2O_5 cell which was based upon an optimized ionic-liquid based electrolyte,

$[Ca(BH_4)_2]_{0.05}[N_{07}TFSI]_{0.95}$, offered an initial discharge capacity of 332mAh/g and a reversible capacity of 244mAh/g.

Mixtures of water and propylene carbonate which contained $Ca(CF_3SO_3)_2$ were used[154] as electrolytes in order to improve the electrochemical performance of a Prussian Blue analogue electrode for use in calcium-ion batteries. The performance was greatly affected by the molar ratio of the water and propylene carbonate in the electrolyte. The electrode had a relatively high capacity when the molar ratio of calcium cations to water in the electrolyte was 1:6. The electrode had a coulombic efficiency of about 100% over 800 cycles, with some 94% retention of the maximum capacity occurring after 800 cycles in this solution. The calcium cations were preferentially solvated by water molecules, and this was held to be related to the improved electrode performance.

Without a suitable combination of electrode materials and electrolyte, it is not possible to ensure a satisfactory performance of complete calcium-ion energy-storage devices. A multi-ion reaction strategy was used[155] to construct a complete calcium-ion energy storage device with a capacitor-battery hybrid mechanism. This device exhibited a reversible capacity of 92mAh/g, a high rate capability and a capacity-retention of 84% over 1000 cycles at room temperature. At the time, this was the best performance reported for a calcium-based energy storage device.

The chemical structure of carbonate electrolyte solvents has a marked influence upon calcium-ion batteries. A computational investigation was made[156] of the solvation behaviour of calcium tetrafluoroborate in both pristine carbonates and in carbonate mixtures solvents and four binary mixtures of those carbonates by using molecular dynamics simulations and quantum mechanical calculations. This showed that both pristine ethyl methyl carbonate and a mixture of ethylene carbonate and diethyl carbonate exhibited the highest free energy of solvation for the Ca^{2+} ion. Interaction of the cation with the carbonyl components of the coordinating solvents, rather than with tetrafluoroborate counter-ions, played the main role in delocalizing the charge on Ca^{2+}. More detailed calculations showed that the HOMO-LUMO energy-gap, the electronic chemical potential and the chemical hardness of the calcium-carbonate complexes were directly proportional to the free energy of solvation of the complex. The results identified pure ethyl methyl carbonate and a binary mixture of ethylene carbonate and diethyl carbonate as being the best electrolytes for calcium-ion batteries. A comparison of the free energy of solvation of the Ca^{2+} ion in the carbonate solvents, with data on Li^+, Na^+ and Mg^{2+} ions, indicated that calcium was far more soluble than the other cations. The Ca^{2+} ions were stabilized by the oxygen atoms of the carbonyl groups and the fluorine atoms of BF_4^- anions. The electrophilic nature of the carbonate solvents increased with Ca^{2+}-ion complexation and with the addition of ethylene carbonate.

Nanotubes can selectively conduct ions across membranes and, depending upon the surface-charge profile of the nanopore, various devices can be constructed. It was shown[157] that a uniformly charged conical nanopore can exhibit many transport properties upon changing the ion species and their concentrations on opposite sides of a membrane. The cation versus anion selectivity of the pores can also be changed. Polyvalent cations such as Ca^{2+} and trivalent cobalt sepulchrate produce a localized charge inversion which changes the effective pore-surface charge-profile from negative to positive. These effects are reversible, thus permitting transport and selectivity characteristics to be tuned.

Calcium-ion batteries can suffer from low electrochemical reversibility, dendrite growth at metal anodes, slow ion-kinetics in metal-oxide cathodes and poor electrode compatibility with organic-based electrolytes. Aqueous multivalent-ion batteries with concentrated aqueous gel electrolytes, sulphur-containing anodes and high-voltage metal-oxide cathodes have been proposed as alternatives to non-aqueous multivalent metal batteries. A room-temperature calcium-ion/sulphur|metal-oxide full cell offered a specific energy of 110Wh/kg and good cycling stability[158]. Molecular dynamics modelling and experimental data indicated that side-reactions could be greatly limited by suppressed water activity and the formation of a protective inorganic solid electrolyte interphase.

A silver sulphide reference electrode has been used[159] for the development of calcium-ion batteries. It consisted of a silver wire which was coated with Ag_2S crystals. Structural characterization was performed by means of Raman spectroscopy and 48h cyclic voltammetry. Three non-aqueous calcium-ion electrolytes were used, and the standard reduction potential of Ag_2S relative to the standard hydrogen electrode was determined, together with the voltage-drift with time. The Ag_2S standard reduction potential ranged from -0.291 to $-0.477V_{SHE}$, and the drift exhibited a linear behaviour, with slopes of between -0.28 and -2.45mV/h. A side-reaction between ferrocene and silver ions occurred, with voltages above 0.65V.

A highly-reversible room-temperature calcium-ion based hybrid battery was created[160] by using a tri-ion strategy which greatly improved the diffusion kinetics of the calcium ions. The optimized battery offered a rate capability of 15C and good cycling stability over 1500 cycles, with 86% capacity-retention at 5C.

A rechargeable Ca/Cl_2 battery was based[161] upon a reversible cathode redox reaction between $CaCl_2$ and Cl_2 which, ironically, was enabled by the use of lithium difluoro(oxalate)borate as a key electrolyte component which facilitated the dissociation and distribution of chlorine-based species and Ca^{2+}. This battery offered discharge

Materials Research Forum LLC
https://doi.org/10.21741/9781644903490

voltages of 3V, together with a specific capacity of 1000mAh/g, a rate capability of 500mA/g and 96.5% capacity-retention after 30 days. It could also operate at temperatures down to 0C.

Upon investigating the materials required for a calcium-ion battery, it was found[162] that an acetonitrile-based solution and a tunnel-structured solid were suitable as the electrolyte and the host material, respectively. Photo-rechargeable air batteries which discharged by reducing atmospheric oxygen and could be charged by incident light were proposed as a secondary battery. In particular, one design for such a photo-recharging air battery employed a metal-hydride anode in a $SrTiO_3LaNi_{3.76}Al_{1.24}H_n|KOH|O_2$ cell. This confirmed that this novel electrode prevented self-discharge but also permitted band-bending in the semiconductor and thus photo-recharging.

Full-cell calcium-ion batteries were made from Prussian Blue and metal-organic compounds[163]. The anode material was created by the precipitation of a nickel salt and various organic sodium salts: $C_6H_{6-x-y}(COONa)_x(NH_2)_y$. There was partial hydrolysis of the nickel-based metal-organic compounds during preparation. The addition of NH_2 groups to the ligand increased the steric hindrance, and partial hydrolysis of the nickel-based metal-organic compounds was inhibited. When the anode material was partially hydrolyzed $Ni(OH)[C_6H_4(COOH)(COO)]$, the full cell exhibited an initial discharge capacity of 82mAh/g and a capacity retention of 62% at 0.1A/g after 100 cycles. The Coulombic efficiency remained at about 95.4%. A cell which was made from NH_2-functionalized $Ni[C_6H_4(NH_2)(COO)_2]$ offered an initial discharge capacity of 86mAh/g and a capacity retention of 77% at 0.1A/g after 100 cycles. The coulombic efficiency remained at about 97%. Impedance tests suggested that the nickel-based metal-organic compounds had a low resistance.

An aqueous calcium-ion battery was based upon a polymerized polyaniline anode and a high redox-potential open-framework structured potassium copper hexacyanoferrate cathode[164]. The charge-discharge mechanism of the battery involved the doping/de-doping of NO_3^- at the anode and intercalation and de-intercalation of calcium ions at the cathode. The battery functioned with a 2.5M $Ca(NO_3)_2$ aqueous electrolyte and offered a specific energy of 70Wh/kg at 250W/kg and retained an energy density of 53Wh/kg at a rate of 950W/kg. At 0.8A/g, the battery offered an average specific capacity of 130mAh/g and exhibited a coulombic efficiency of about 96%, with 95% capacity retention after 200 cycles.

An aqueous calcium-ion battery was based upon mesoporous silica with a 2-dimensional hexagonal through-hole structure[165]. An organic material, supported on the silica, was used as the anode and this led to a capacity of 201mAh/g, together with a stable cycling behaviour and 95% capacity-retention after 1500 cycles. When combined with Ca_2MnO_4, the aqueous battery offered an energy-density of 130.6Wh/kg with a voltage range of 0 to 1.8V, together with a high capacity and good cycling stability. The silica anode material provided a particular bonding of redox electrons, and this led to the highly stable behaviour. This in turn solved the problems of existing organic electrode materials. Localization and de-localization of the redox electrons led to additional voltage stability.

With the aim of finding suitable electrolyte salts for rechargeable calcium-ion batteries, complexes of $Ca(PF_6)_2$, a key potential component of such electrolytes, was prepared[166] by treating calcium metal with $NOPF_6$, and its conversion to species containing $PO_2F_2^-$.

The possibility of Ca^{2+} intercalation into a layered Na_2FePO_4F host was studied[167]. This was the first example of Ca^{2+} intercalation into a polyanionic framework and the results suggested that other polyanionic framework materials could be suitable for Ca^{2+} intercalation.

About the Author

Dr. Fisher has wide knowledge and experience of the fields of engineering, metallurgy and solid-state physics, beginning with work at Rolls-Royce Aero Engines on turbine-blade research, related to the Concord supersonic passenger-aircraft project, which led to a BSc degree (1971) from the University of Wales. This was followed by theoretical and experimental work on the directional solidification of eutectic alloys having the ultimate aim of developing composite turbine blades. This work led to a doctoral degree (1978) from the Swiss Federal Institute of Technology (Lausanne). He then acted for many years as an editor of various academic journals, in particular *Defect and Diffusion Forum*. In recent years he has specialized in writing monographs which introduce readers to the most rapidly developing ideas in the fields of engineering, metallurgy and solid-state physics. He is co-author of the widely-cited student textbook, *Fundamentals of Solidification*. Google Scholar credits him with 9614 citations and a lifetime h-index of 13.

References

[1] Wu S., Zhang F., Tang Y., Advanced Science, 5[8] 2018, 1701082.
https://doi.org/10.1002/advs.201701082

[2] Wang M., Jiang C., Zhang S., Song X., Tang Y., Cheng H.M., Nature Chemistry,
10[6] 2018, 667-672. https://doi.org/10.1038/s41557-018-0045-4

[3] Prabakar S.J.R., Ikhe A.B., Park W.B., Chung K.C., Park H., Kim K.J., Ahn D.,
Kwak J.S., Sohn K.S., Pyo M., Advanced Science, 6[24] 2019, 1902129.

[4] Park J., Xu Z.L., Yoon G., Park S.K., Wang J., Hyun H., Park H., Lim J., Ko Y.J.,
Yun Y.S., Kang K., Advanced Materials, 32[4] 2020, 1904411.
https://doi.org/10.1002/adma.201904411

[5] Nishimura Y., Nakatani N., Nakagawa K., Journal of Solid State Electrochemistry,
25[10-11] 2021, 2495-2501. https://doi.org/10.1007/s10008-021-05024-7

[6] Kwak J.H., Hyun J.C., Park J.H., Chung K.Y., Yu S.H., Yun Y.S., Lim H.D., Journal
of Industrial and Engineering Chemistry, 96, 2021, 397-403.
https://doi.org/10.1016/j.jiec.2021.02.006

[7] Li J., Han C., Ou X., Tang Y., Angewandte Chemie, 61[14] 2022, e202116668.
https://doi.org/10.1002/anie.202116668

[8] Nishimura Y., Yamazaki S., Sakoda T., Nakagawa K., SN Applied Sciences, 5[2]
2023, 58. https://doi.org/10.1007/s42452-023-05275-1

[9] Yi Y., Xing Y., Wang H., Zeng Z., Sun Z., Li R., Lin H., Ma Y., Pu X., Li M.M.J.,
Park K.Y., Xu Z.L., Angewandte Chemie, 63[24] 2024, e202317177.
https://doi.org/10.1002/anie.202317177

[10] Niaei A.H.F., Hussain T., Hankel M., Searles D.J., Carbon, 136, 2018, 73-84.
https://doi.org/10.1016/j.carbon.2018.04.034

[11] Gao Y., Li Z., Wang P., Cui W., Wang X., Yang Y., Gao F., Zhang M., Gan J., Li
C., Liu Y., Wang X., Qi F., Zhang J., Han X., Du W., Pan H., Xia Z., Advanced
Functional Materials, 33[50] 2023, 2305610. https://doi.org/10.1002/adfm.202305610

[12] Hassanpour A., Farhami N., Derakhshande M., Nezhad P.D.K., Ebadi A.,
Ebrahimiasl S., Inorganic Chemistry Communications, 129, 2021, 108656.
https://doi.org/10.1016/j.inoche.2021.108656

[13] Buenaño L., Ajaj Y., Padilla C., Barahona B.V., Mejía N., Real R.B.D., Oviedo
B.S.R., Fiallos J.J.F., Saraswat S.K., Chemical Physics, 579, 2024, 112194.
https://doi.org/10.1016/j.chemphys.2024.112194

[14] Yang T., Ma T.C., Ye X.J., Zheng X.H., Jia R., Yan X.H., Liu C.S., Physical
Chemistry Chemical Physics, 26[5] 2024, 4589-4596.
https://doi.org/10.1039/D3CP04976K

[15] Linganay J.B.D., Putungan D.B., Materials Research Express, 9[9] 2022, 095506. https://doi.org/10.1088/2053-1591/ac92c8

[16] Zhang X., Li F., Wu P., Wang Y., Zhang J., Xie Y., Zhu Y., Zhu X., Wei S., Zhou Y., Journal of Colloid and Interface Science, 545, 2019, 200-208. https://doi.org/10.1016/j.jcis.2019.03.029

[17] Keshtkari L., Rabczuk T., Physica E, 161, 2024, 115955. https://doi.org/10.1016/j.physe.2024.115955

[18] Adil M., Dutta P.K., Mitra S., ChemistrySelect, 3[13] 2018, 3687-3690. https://doi.org/10.1002/slct.201800419

[19] Zhang S., Zhu Y., Wang D., Li C., Han Y., Shi Z., Feng S., Advanced Science, 9[14] 2022, 2200397. https://doi.org/10.1002/advs.202200397

[20] Abduryim E., Chen C., Gao L., Guo S., Wang S., Zhang Z., Cai Y., Gao S., Chen W., Guan X., Liu Y., Liu G., Lu P., Journal of Physical Chemistry C, 128[15] 2024, 6233-6248. https://doi.org/10.1021/acs.jpcc.4c00557

[21] Liu Y., He B., Pu J., Yu M., Zhang Y., Meng C., Zhang Q., Wu J., Wei L., Pan Z., Journal of Energy Chemistry, 99, 2024, 661-670. https://doi.org/10.1016/j.jechem.2024.08.014

[22] Jiang B., Su Y., Liu R., Sun Z., Wu D., Small, 18[20] 2022, 2200049. https://doi.org/10.1002/smll.202200049

[23] Li L., Zhang G., Deng X., Hao J., Zhao X., Li H., Han C., Li B., Journal of Materials Chemistry A, 10[39] 2022, 20827-20836. https://doi.org/10.1039/D2TA05185K

[24] Zhang S., Zhu Y.L., Ren S., Li C., Chen X.B., Li Z., Han Y., Shi Z., Feng S., Journal of the American Chemical Society, 145[31] 2023, 17309-17320. https://doi.org/10.1021/jacs.3c04657

[25] Qiao F., Wang J., Yu R., Huang M., Zhang L., Yang W., Wang H., Wu J., Jiang Y., An Q., ACS Nano, 17[22] 2023, 23046-23056. https://doi.org/10.1021/acsnano.3c08645

[26] Li R., Yu J., Chen F., Su Y., Chan K.C., Xu Z.L., Advanced Functional Materials, 33[30] 2023, 2214304. https://doi.org/10.1002/adfm.202214304

[27] Wang C., Li R., Zhu Y., Wang Y., Lin Y., Zhong L., Chen H., Tang Z., Li H., Liu F., Zhi C., Lv H., Advanced Energy Materials, 14[1] 2024, 2302495. https://doi.org/10.1002/aenm.202302495

[28] Das P., Ball B., Sarkar P., Physical Chemistry Chemical Physics, 24[36], 2022, 21729-21739. https://doi.org/10.1039/D2CP02852B

[29] Maltsev A.P., Chepkasov I.V., Oganov A.R., Materials Today Energy, 39, 2024, 101467. https://doi.org/10.1016/j.mtener.2023.101467

[30] Jao W.Y., Tai C.W., Chang C.C., Hu C.C., Energy Storage Materials, 63, 2023, 102990. https://doi.org/10.1016/j.ensm.2023.102990

[31] Ahmed S., Ghani A., Muhammad I., Muhammad I., Mehmood A., Ullah N., Hassan A., Wang Y., Tian X., Yakobson B., Physical Chemistry Chemical Physics, 26[8] 2024, 6977-6983. https://doi.org/10.1039/D3CP05171D

[32] Han C., Li H., Li Y., Zhu J., Zhi C., Nature Communications, 12[1] 2021, 2400. https://doi.org/10.1038/s41467-021-22698-9

[33] Cao Y., Sharma K., Rajhi A.A., Alamri S., Anqi A.E., El-Shafay A.S., Aly A.A., Felemban B.F., Rashidi S., Derakhshandeh M., Journal of Electroanalytical Chemistry, 910, 2022, 115929. https://doi.org/10.1016/j.jelechem.2021.115929

[34] Putungan D.B., Llemit C.L.T., Santos-Putungan A.B., Sarmago R.V., Gebauer R., Physical Chemistry Chemical Physics, 26[5] 2024, 4298-4305. https://doi.org/10.1039/D3CP04897G

[35] Kadhim M.M., Rheima A.M., Shadhar M.H., Abdulnabi S.M., Saleh Z.M., Al Mashhadani Z.I., Najm Z.M., Sarkar A., Silicon, 15[1] 2023, 417-424. https://doi.org/10.1007/s12633-022-02014-w

[36] Park J., Fatima S.A., Journal of Energy Storage, 98, 2024, 113111. https://doi.org/10.1016/j.est.2024.113111

[37] Shiga T., Kondo H., Kato Y., Inoue M., Journal of Physical Chemistry C, 119[50], 2015, 27946-27953. https://doi.org/10.1021/acs.jpcc.5b10245

[38] Lee H., Kim D., Jeong S.K., Key Engineering Materials, 803, 2019, 109-114. https://doi.org/10.4028/www.scientific.net/KEM.803.109

[39] Padigi P., Goncher G., Evans D., Solanki R., Journal of Power Sources, 273, 2015, 460-464. https://doi.org/10.1016/j.jpowsour.2014.09.101

[40] Tojo T., Sugiura Y., Inada R., Sakurai Y., Electrochimica Acta, 207, 2016, 22-27. https://doi.org/10.1016/j.electacta.2016.04.159

[41] Kuperman N., Padigi P., Goncher G., Evans D., Thiebes J., Solanki R., Journal of Power Sources, 342, 2017, 414-418. https://doi.org/10.1016/j.jpowsour.2016.12.074

[42] Lee C., Jeong S.K., Electrochimica Acta, 265, 2018, 430-436. https://doi.org/10.1016/j.electacta.2018.01.172

[43] Du C.Y., Zhang Z.H., Li X.L., Luo R.J., Ma C., Bao J., Zeng J., Xu X., Wang F., Zhou Y.N., Chemical Engineering Journal, 451, 2023, 138650. https://doi.org/10.1016/j.cej.2022.138650

[44] Vo T.N., Kang J.E., Lee H., Lee S.W., Ahn S.K., Hur J., Kim I.T., EcoMat, 5[2] 2023, e12285. https://doi.org/10.1002/eom2.12285

[45] Kadhim M.M., Shadhar M.H., Nathir I., Taban T.Z., Noori A.S., Almashhadani H.A., Rheima A.M., Ebadi A.G., International Journal of Hydrogen Energy, 47[74] 2022, 31665-31672. https://doi.org/10.1016/j.ijhydene.2022.07.087

[46] Xiong Y., Ma N., Wang Y., Wang T., Luo S., Fan J., Physical Chemistry Chemical Physics, 25[18] 2023, 12854-12862. https://doi.org/10.1039/D3CP00352C

[47] Hadi M.A., Kadhim M.M., Al-Azawi I.I.K., Abdullaha S.A.H., Majdi A., Hachim S.K., Rheima A.M., Computational and Theoretical Chemistry, 1219, 2023, 113940. https://doi.org/10.1016/j.comptc.2022.113940

[48] Jaber N.A., Abed Z.T., Kadhim M.M., Yaseen Y., Khazaal W.M., Almashhadani H.A., Rheima A.M., Mohamadi A., Materials Chemistry and Physics, 294, 2023, 126926. https://doi.org/10.1016/j.matchemphys.2022.126926

[49] Chen Y., Bartel C.J., Avdeev M., Zhang Y.Q., Liu J., Zhong P., Zeng G., Cai Z., Kim H., Ji H., Ceder G., Chemistry of Materials, 34[1] 2022, 128-139. https://doi.org/10.1021/acs.chemmater.1c02923

[50] Hussain T., Searles D.J., Hankel M., Carbon, 160, 2020, 125-132. https://doi.org/10.1016/j.carbon.2019.12.063

[51] Saharan S., Ghanekar U., Shivankar B.R., Meena S., Journal of Physical Chemistry C, 128[31] 2024, 12840-12848. https://doi.org/10.1021/acs.jpcc.4c03063

[52] Çiftçi N.O., Sentürk S.B., Sezen Y., Kaykusuz S.Ü., Long H., Ergen O., Proceedings of the National Academy of Sciences of the United States of America, 120[42] 2023, e2307537120. https://doi.org/10.1073/pnas.2307537120

[53] Li X., Li Y., Wang Y., Journal of Molecular Modeling, 29[10] 2023, 308. https://doi.org/10.1007/s00894-023-05714-1

[54] Wang Y., Liang S., Tian J., Duan H., Lv Y., Wan L., Huang C., Wu M., Ouyang C., Hu J., Physical Chemistry Chemical Physics, 26[5] 2024, 4455-4465. https://doi.org/10.1039/D3CP05287G

[55] Masood M.K., Wang J., Song J., Liu Y., Journal of Materials Chemistry A, 12[34] 2024, 22945-22959. https://doi.org/10.1039/D4TA02176B

[56] Zhou R., Hou Z., Liu Q., Du X., Huang J., Zhang B., Advanced Functional Materials, 32[26] 2022, 2200929. https://doi.org/10.1002/adfm.202200929

[57] Kim S., Hahn N.T., Fister T.T., Leon N.J., Lin X.M., Park H., Zapol P., Lapidus S.H., Liao C., Vaughey J.T., Chemistry of Materials, 35[6] 2023, 2363-2370. https://doi.org/10.1021/acs.chemmater.2c03343

[58] Chen K., Feng Y., International Journal of Quantum Chemistry, 124[15] 2024, e27457. https://doi.org/10.1002/qua.27448

[59] Niu X., Feng Y., Journal of Molecular Modeling, 30[7] 2024, 211. https://doi.org/10.1007/s00894-024-06011-1

[60] Ren W., Xiong F., Fan Y., Xiong Y., Jian Z., ACS Applied Materials and Interfaces, 12[9] 2020, 10471-10478. https://doi.org/10.1021/acsami.9b21999

[61] Chowdhury S., Sen P., Gupta B.C., Computational Materials Science, 230, 2023, 112539. https://doi.org/10.1016/j.commatsci.2023.112539

[62] Seong S., Lee H., Lee S., Nogales P.M., Lee C., Kim Y., Jeong S.K., Batteries, 9[10] 2023, 500. https://doi.org/10.3390/batteries9100500

[63] Lee C., Jeong Y.T., Nogales P.M., Song H.Y., Kim Y., Yin R.Z., Jeong S.K., Electrochemistry Communications, 98, 2019, 115-118. https://doi.org/10.1016/j.elecom.2018.12.003

[64] Pan H., Wang C., Qiu M., Wang Y., Han C., Nan D., Energy and Environmental Materials, 7[5] 2024, e12690. https://doi.org/10.1002/eem2.12690

[65] Fu T., Feng Y., Gao W., Li X., Journal of Molecular Modeling, 30[2], 2024, 34. https://doi.org/10.1007/s00894-024-05829-z

[66] Liu N., Feng Y., Li X., Yu W., Journal of Molecular Modeling, 30[5] 2024, 119. https://doi.org/10.1007/s00894-024-05904-5

[67] Alabada R., Moreano G., Guamán A., Yadav A., Barahona M., Viñan J., Makrariya A., J. Saadh M., Elmasry Y., Inorganic Chemistry Communications, 159, 2024, 111801. https://doi.org/10.1016/j.inoche.2023.111801

[68] Kim S., Yin L., Lee M.H., Parajuli P., Blanc L., Fister T.T., Park H., Kwon B.J., Ingram B.J., Zapol P., Klie R.F., Kang K., Nazar L.F., Lapidus S.H., Vaughey J.T., ACS Energy Letters, 5[10] 2020, 3203-3211. https://doi.org/10.1021/acsenergylett.0c01663

[69] Li R., Lee Y., Lin H., Che X., Pu X., Yi Y., Chen F., Yu J., Chan K.C., Park K.Y., Xu Z.L., Advanced Energy Materials, 14[11] 2024, 2302700. https://doi.org/10.1002/aenm.202302700

[70] Qiao F., Wang J., Yu R., Pi Y., Huang M., Cui L., Liu Z., An Q., Small Methods, 8[1] 2024, 2300865. https://doi.org/10.1002/smtd.202300865

[71] Xu Z.L., Park J., Wang J., Moon H., Yoon G., Lim J., Ko Y.J., Cho S.P., Lee S.Y., Kang K., Nature Communications, 12[1] 2021, 3369. https://doi.org/10.1038/s41467-021-23703-x

[72] Jeon B., Heo J.W., Hyoung J., Kwak H.H., Lee D.M., Hong S.T., Chemistry of Materials, 32[20] 2020, 8772-8780. https://doi.org/10.1021/acs.chemmater.0c01112

[73] Wang J., Tan S., Xiong F., Yu R., Wu P., Cui L., An Q., Chemical Communications, 56[26] 2020, 3805-3808. https://doi.org/10.1039/D0CC00772B

[74] Prabakar S.J.R., Park W.B., Seo J.Y., Singh S.P., Ahn D., Sohn K.S., Pyo M., Energy Storage Materials, 43, 2021, 85-96. https://doi.org/10.1016/j.ensm.2021.08.035

[75] Ponrouch A., Tchitchekova D., Frontera C., Bardé F., Dompablo M.E.A.D., Palacín M.R., Electrochemistry Communications, 66, 2016, 75-78. https://doi.org/10.1016/j.elecom.2016.03.004

[76] Muhammad I., Ahmed S., Cao H., Mahmood A., Wang Y.G., Journal of Physical Chemistry C, 127[2] 2023, 1198-1208. https://doi.org/10.1021/acs.jpcc.2c06877

[77] Sharma A., Thomas A., Gupta P., Batteries and Supercaps, 6[8] 2023, e202300073. https://doi.org/10.1002/batt.202300073

[78] Kadhim M.M., Majdi A., Hachim S.K., Abdullaha S.A., Taban T.Z., Rheima A.M., Korean Journal of Chemical Engineering, 40[7] 2023, 1633-1638. https://doi.org/10.1007/s11814-023-1433-z

[79] Imaduddin I.S., Idris N.H., Majid S.R., Journal of Energy Storage, 94, 2024, 112362. https://doi.org/10.1016/j.est.2024.112362

[80] Cabello M., Nacimiento F., González J.R., Ortiz G., Alcántara R., Lavela P., Pérez-Vicente C., Tirado J.L., Electrochemistry Communications, 67, 2016, 59-64. https://doi.org/10.1016/j.elecom.2016.03.016

[81] Kuganathan N., Ganeshalingam S., Chroneos A., AIP Advances, 10[7] 2020, 075004. https://doi.org/10.1063/5.0012594

[82] Dompablo M.E.A.D., Krich C., Nava-Avendaño J., Biškup N., Palacín M.R., Bardé F., Chemistry of Materials, 28[19] 2016, 6886-6893. https://doi.org/10.1021/acs.chemmater.6b02146

[83] Chando P.A., Chen S., Shellhamer J.M., Wall E., Wang X., Schuarca R., Smeu M., Hosein I.D., Chemistry of Materials, 35[20] 2023, 8371-8381. https://doi.org/10.1021/acs.chemmater.3c00659

[84] Hyoung J., Heo J.W., Hong S.T., Journal of Power Sources, 390, 2018, 127-133. https://doi.org/10.1016/j.jpowsour.2018.04.050

[85] Shi Z., Wu J., Ni M., Guo Q., Zan F., Xia H., Materials Research Bulletin, 144, 2021, 111475. https://doi.org/10.1016/j.materresbull.2021.111475

[86] Xu F., Shi Z., Wu J., Liu H., Li J., Zan F., Xia H., Journal of Power Sources, 602, 2024, 234342. https://doi.org/10.1016/j.jpowsour.2024.234342

[87] Zuo C., Xiong F., Wang J., An Y., Zhang L., An Q., Advanced Functional Materials, 32[33] 2022, 2202975. https://doi.org/10.1002/adfm.202202975

[88] Tojo T., Tawa H., Oshida N., Inada R., Sakurai Y., Journal of Electroanalytical Chemistry, 825, 2018, 51-56. https://doi.org/10.1016/j.jelechem.2018.08.008

[89] Cabello M., Nacimiento F., Alcántara R., Lavela P., Pérez Vicente C., Tirado J.L., Chemistry of Materials, 30[17], 2018, 5853-5861. https://doi.org/10.1021/acs.chemmater.8b01116

[90] Chae M.S., Kwak H.H., Hong S.T., ACS Applied Energy Materials, 3[6] 2020, 5107-5112. https://doi.org/10.1021/acsaem.0c00567

[91] Kim S., Yin L., Bak S.M., Fister T.T., Park H., Parajuli P., Gim J., Yang Z., Klie R.F., Zapol P., Du Y., Lapidus S.H., Vaughey J.T., Nano Letters, 22[6] 2022, 2228-2235. https://doi.org/10.1021/acs.nanolett.1c04157

[92] Qin Z., Song Y., Liu Y., Liu X.X., Chemical Engineering Journal, 451, 2022, 138681. https://doi.org/10.1016/j.cej.2022.138681

[93] Zuo C., Shao Y., Li M., Zhang W., Zhu D., Tang W., Hu J., Liu P., Xiong F., An Q., ACS Applied Materials and Interfaces , 16[26] 2024, 33733-33739. https://doi.org/10.1021/acsami.4c07585

[94] Park H., Bartel C.J., Ceder G., Zapol P., Advanced Energy Materials, 11[48] 2021, 2101698. https://doi.org/10.1002/aenm.202101698

[95] Wang J., Zhang Y., Qiao F., Jiang Y., Yu R., Li J., Lee S., Dai Y., Guo F., Jiang P., Zhang L., An Q., He G., Mai L., Advanced Materials, 36[30] 2024, 2403371. https://doi.org/10.1002/adma.202403371

[96] Wang J., Yu R., Jiang Y., Qiao F., Liao X., Wang J., Huang M., Xiong F., Cui L., Dai Y., Zhang L., An Q., He G., Mai L., Energy and Environmental Science, 17, 2024, 6616-6626. https://doi.org/10.1039/D4EE02003K

[97] Xiang L., Yang W., Wang Y., Sun X., Xu J., Cao D., Li Q., Li H., Wang X., Chemical Communications, 60[41] 2024, 5459-5462. https://doi.org/10.1039/D4CC00988F

[98] Tekliye D.B., Kumar A., Weihang X., Mercy T.D., Canepa P., Sai Gautam G., Chemistry of Materials, 34[22] 2022, 10133-10143. https://doi.org/10.1021/acs.chemmater.2c02841

[99] Wang D., Liu H., Elliott J.D., Liu L.M., Lau W.M., Journal of Materials Chemistry A, 4[32] 2016, 12516-12525. https://doi.org/10.1039/C6TA03595G

[100] Murata Y., Takada S., Obata T., Tojo T., Inada R., Sakurai Y., Electrochimica Acta, 294, 2019, 210-216. https://doi.org/10.1016/j.electacta.2018.10.103

[101] Xu X., Duan M., Yue Y., Li Q., Zhang X., Wu L., Wu P., Song B., Mai L., ACS Energy Letters, 4[6] 2019, 1328-1335. https://doi.org/10.1021/acsenergylett.9b00830

[102] Chae M.S., Heo J.W., Hyoung J., Hong S.T., ChemNanoMat, 6[7] 2020, 1049-1053. https://doi.org/10.1002/cnma.202000011

[103] Jeon B., Kwak H.H., Hong S.T., Chemistry of Materials, 34[4] 2022, 1491-1498. https://doi.org/10.1021/acs.chemmater.1c02774

[104] Prabakar S.J.R., Ikhe A.B., Park W.B., Ahn D., Sohn K.S., Pyo M., Advanced Functional Materials, 33[29] 2023, 2301399.

[105] Hu Z., Chang S., Cheng C., Sun C., Liu J., Meng T., Xuan Y., Ni M., Energy Storage Materials, 58, 2023, 353-361. https://doi.org/10.1016/j.ensm.2023.03.037

[106] Lakhnot A.S., Bhimani K., Mahajani V., Panchal R.A., Sharma S., Koratkar N., Proceedings of the National Academy of Sciences of the United States of America, 119[30] 2022, e2205762119. https://doi.org/10.1073/pnas.2205762119

[107] Zhao X., Li L., Zheng L., Fan L., Yi Y., Zhang G., Han C., Li B., Advanced Functional Materials, 34[2] 2024, 2309753. https://doi.org/10.1002/adfm.202309753

[108] Qin X., Zhao X., Zhang G., Wei Z., Li L., Wang X., Zhi C., Li H., Han C., Li B., ACS Nano, 17[13] 2023, 12040-12051. https://doi.org/10.1021/acsnano.2c07061

[109] Tan X., Wang J., Jin S., Wang Y., Qiao F., Zhang L., An Q., New Journal of Chemistry, 47[17] 2023, 8326-8333. https://doi.org/10.1039/D3NJ00886J

[110] Dong L.W., Xu R.G., Wang P.P., Sun S.C., Li Y., Zhen L., Xu C.Y., Journal of Power Sources, 479, 2020, 228793. https://doi.org/10.1016/j.jpowsour.2020.228793

[111] Zeng F., Li S., Hu S., Qiu M., Zhang G., Li M., Chang C., Wang H., Xu M., Zheng L., Tang Y., Han C., Cheng H.M., Advanced Functional Materials, 34[5] 2024, 2302397. https://doi.org/10.1002/adfm.202302397

[112] Zhao X., Li L., Zhang G., Yi Y., Yang T., Han C., Li B., Small Methods, 8[6] 2024, 2400097. https://doi.org/10.1002/smtd.202400097

[113] Luo P., Zhu D., Zuo C., Zhang W., Li M., Chao F., Yu G., Zhong W., An Q., Chemical Engineering Journal, 482, 2024, 149177. https://doi.org/10.1016/j.cej.2024.149177

[114] Hao X., Zheng L., Hu S., Wu Y., Zhang G., Li B., Yang M., Han C., Materials Today Energy, 38, 2023, 101456. https://doi.org/10.1016/j.mtener.2023.101456

[115] Feng H., Wang Y., Qiu W., Liu Z., Tao Y., Lu X., Inorganic Chemistry Frontiers, 11[11] 2024, 3136-3149. https://doi.org/10.1039/D4QI00218K

[116] Wang C., Wang J., Zhang S., Li M., Zeng F., Tan L., Liu F., Wang J., Huang L., Lv H., Zhi C., Han C., Advanced Energy Materials, 13[41] 2023, 2302683. https://doi.org/10.1002/aenm.202302683

[117] Wang C.F., Zhang S.W., Huang L., Zhu Y.M., Liu F., Wang J.C., Tan L.M., Zhi C.Y., Han C.P., Rare Metals, 43[6] 2024, 2597-2612. https://doi.org/10.1007/s12598-023-02613-5

[118] Chae M.S., Setiawan D., Kim H.J., Hong S.T., Batteries, 7[3] 2021, 54.
https://doi.org/10.3390/batteries7030054

[119] Liu L., Wu Y.C., Rozier P., Taberna P.L., Simon P., Research, 2019, 6585686.
https://doi.org/10.34133/2019/6585686

[120] Wang J., Wang J., Jiang Y., Xiong F., Tan S., Qiao F., Chen J., An Q., Mai L.,
Advanced Functional Materials, 32[25] 2022, 2113030.
https://doi.org/10.1002/adfm.202113030

[121] Purbarani M.E., Hyoung J., Hong S.T., ACS Applied Energy Materials, 4[8] 2021,
7487-7491. https://doi.org/10.1021/acsaem.1c01158

[122] Hyoung J., Heo J.W., Jeon B., Hong S.T., Journal of Materials Chemistry A, 9[36]
2021, 20776-20782. https://doi.org/10.1039/D1TA03881H

[123] Vo T.N., Kim H., Hur J., Choi W., Kim I.T., Journal of Materials Chemistry A,
6[45] 2018, 22645-22654. https://doi.org/10.1039/C8TA07831A

[124] Adil M., Sarkar A., Sau S., Muthuraj D., Mitra S., Journal of Power Sources, 541,
2022, 231669. https://doi.org/10.1016/j.jpowsour.2022.231669

[125] Bu H., Lee H., Hyoung J., Heo J.W., Kim D., Lee Y.J., Hong S.T., Chemistry of
Materials, 35[19] 2023, 7974-7983. https://doi.org/10.1021/acs.chemmater.3c01207

[126] Wang D., Gao X., Chen Y., Jin L., Kuss C., Bruce P.G., Nature Materials, 17[1]
2018, 16-20. https://doi.org/10.1038/nmat5036

[127] Liu H.Y., Wang Z.Y., Colloids and Surfaces A, 692, 2024, 133886.
https://doi.org/10.1016/j.colsurfa.2024.133928

[128] Young J., Smeu M., Advanced Theory and Simulations, 4[8] 2021, 2100018.
https://doi.org/10.1002/adts.202100018

[129] Zhao-Karger Z., Xiu Y., Li Z., Reupert A., Smok T., Fichtner M., Nature
Communications, 13[1] 2022, 3849. https://doi.org/10.1038/s41467-022-31261-z

[130] Yao Z., Hegde V.I., Aspuru-Guzik A., Wolverton C., Advanced Energy Materials,
9[9] 2019, 1802994. https://doi.org/10.1002/aenm.201802994

[131] Jie Y., Tan Y., Li L., Han Y., Xu S., Zhao Z., Cao R., Ren X., Huang F., Lei Z.,
Tao G., Zhang G., Jiao S., Angewandte Chemie, 59[31] 2020, 12689-12693.
https://doi.org/10.1002/anie.202002274

[132] Jeong H., Kamphaus E.P., Redfern P.C., Hahn N.T., Leon N.J., Liao C., Cheng L.,
ACS Applied Materials and Interfaces, 15[5] 2023, 6933-6941.
https://doi.org/10.1021/acsami.2c20661

[133] Yamijala S.S.R.K.C., Kwon H., Guo J., Wong B.M., ACS Applied Materials and
Interfaces, 13[11] 2021, 13114-13122. https://doi.org/10.1021/acsami.0c21716

[134] Ye C., Li H., Chen Y., Hao J., Liu J., Shan J., Qiao S.Z., Nature Communications, 15[1] 2024, 4797. https://doi.org/10.1038/s41467-024-49164-6

[135] Batzinger K., Liepinya D., Smeu M., Physical Chemistry Chemical Physics, 26[6] 2024, 5218-5225. https://doi.org/10.1039/D3CP04113A

[136] Sharifi S., Asenjo-Sanz I., Verde-Sesto E., Maiz J., Pomposo J.A., Alegría A., Polymer, 287, 2023, 126422. https://doi.org/10.1016/j.polymer.2023.126422

[137] Yang F., Feng X., Zhuo Z., Vallez L., Liu Y.S., McClary S.A., Hahn N.T., Glans P.A., Zavadil K.R., Guo J., Arabian Journal for Science and Engineering, 48[6] 2023, 7243-7262. https://doi.org/10.1007/s13369-022-07597-5

[138] Wang J., Genier F.S., Li H., Biria S., Hosein I.D., ACS Applied Polymer Materials, 1[7] 2019, 1837-1844. https://doi.org/10.1021/acsapm.9b00371

[139] Gerbec J.A., Jun Y.S., Wudl F., Stucky G.D., Seshadri R., Advanced Energy Materials, 3, 2013, 1056-1061. https://doi.org/10.1002/aenm.201300160

[140] Biria S., Pathreeker S., Genier F.S., Hosein I.D., ACS Applied Polymer Materials, 2[6] 2020, 2111-2118. https://doi.org/10.1021/acsapm.9b01223

[141] Fluker E.C., Pathreeker S., Hosein I.D., Journal of Physical Chemistry C, 127[33] 2023, 16579-16587. https://doi.org/10.1021/acs.jpcc.3c02342

[142] Liu Y., Shao Z., Lv T., Zhang Z., Zhou Z., Hu T., Meng C., Zhang Y., Green Energy and Environment, 2024, in press.

[143] Biria S., Pathreeker S., Genier F.S., Chen F.H., Li H., Burdin C.V., Hosein I.D., ACS Omega, 6[26] 2021, 17095-17102. https://doi.org/10.1021/acsomega.1c02312

[144] Pathreeker S., Hosein I.D., ACS Applied Polymer Materials, 4[10] 2022, 6803-6811. https://doi.org/10.1021/acsapm.2c01140

[145] Genier F.S., Burdin C.V., Biria S., Hosein I.D., Journal of Power Sources, 414, 2019, 302-307. https://doi.org/10.1016/j.jpowsour.2019.01.017

[146] Lee C., Jeong S.K., Chemistry Letters, 45[12] 2016, 1447-1449. https://doi.org/10.1246/cl.160769

[147] Adil M., Ghosh A., Mitra S., ACS Applied Materials and Interfaces, 14[22], 2022, 25501-25515. https://doi.org/10.1021/acsami.2c04742

[148] Anh C.D., Kim Y.J., Vo T.N., Kim D., Hur J., Khani H., Kim I.T., Chemical Engineering Journal, 471, 2023, 144631. https://doi.org/10.1016/j.cej.2023.144631

[149] Lu J., Jaumaux P., Wang T., Wang C., Wang G., Journal of Materials Chemistry A, 9[43] 2021, 24175-24194. https://doi.org/10.1039/D1TA06606D

[150] Leon N.J., Ilic S., Xie X., Jeong H., Yang Z., Wang B., Spotte-Smith E.W.C., Stern C., Hahn N., Zavadil K., Cheng L., Persson K.A., Connell J.G., Liao C., Journal of

Physical Chemistry Letters, 15[19] 2024, 5096-5102.
https://doi.org/10.1021/acs.jpclett.4c00969

[151] Karatrantos A.V., Middendorf M., Nosov D.R., Cai Q., Westermann S., Hoffmann K., Nurnberg P., Shaplov A.S., Schonhoff M., Journal of Chemical Physics, 161[5] 2024, 054502. https://doi.org/10.1063/5.0216222

[152] Batzinger K., Smeu M., Physical Chemistry Chemical Physics, 24[48] 2022, 29579-29585. https://doi.org/10.1039/D2CP02274E

[153] Gao X., Liu X., Mariani A., Elia G.A., Lechner M., Streb C., Passerini S., Energy and Environmental Science, 13[8] 2020, 2559-2569.
https://doi.org/10.1039/D0EE00831A

[154] Lee C., Jeong S.K., Electrochemistry, 86[3] 2018, 134-137.
https://doi.org/10.5796/electrochemistry.17-00069

[155] Wu N., Yao W., Song X., Zhang G., Chen B., Yang J., Tang Y., Advanced Energy Materials, 9[16] 2019, 1803865. https://doi.org/10.1002/aenm.201803865

[156] Shakourian-Fard M., Kamath G., Taimoory S.M., Trant J.F., Journal of Physical Chemistry C, 123[26] 2019, 15885-15896. https://doi.org/10.1021/acs.jpcc.9b01655

[157] He Y., Gillespie D., Boda D., Vlassiouk I., Eisenberg R.S., Siwy Z.S., Journal of the American Chemical Society, 131[14] 2009, 5194-5202.
https://doi.org/10.1021/ja808717u

[158] Tang X., Zhou D., Zhang B., Wang S., Li P., Liu H., Guo X., Jaumaux P., Gao X., Fu Y., Wang C., Wang C., Wang G., Nature Communications, 12[1] 2021, 2857.
https://doi.org/10.1038/s41467-021-23209-6

[159] Shellhamer J.M., Chando P.A., Pathreeker S., Wang X., Hosein I.D., Journal of Physical Chemistry C, 127[40] 2023, 19900-19905.
https://doi.org/10.1021/acs.jpcc.3c04580

[160] Lang J., Jiang C., Fang Y., Shi L., Miao S., Tang Y., Advanced Energy Materials, 9[29] 2019, 1901099. https://doi.org/10.1002/aenm.201901099

[161] Geng S., Zhao X., Xu Q., Yuan B., Wang Y., Liao M., Ye L., Wang S., Ouyang Z., Wu L., Wang Y., Ma C., Zhao X., Sun H., Nature Communications, 15[1] 2024, 944.
https://doi.org/10.1038/s41467-024-45347-3

[162] Akuto K., Ohtsuka H., Hayashi M., Nemoto Y., NTT R and D, 50[8] 2001, 592-598.

[163] Vo T.N., Hur J., Kim I.T., ACS Sustainable Chemistry and Engineering, 8[7] 2020, 2596-2601. https://doi.org/10.1021/acssuschemeng.9b07374

[164] Adil M., Sarkar A., Roy A., Panda M.R., Nagendra A., Mitra S., ACS Applied Materials and Interfaces, 12[10] 2020, 11489-11503.
https://doi.org/10.1021/acsami.9b20129

[165] Cang R., Zhao C., Ye K., Yin J., Zhu K., Yan J., Wang G., Cao D., ChemSusChem, 13[15] 2020, 3911-3918. https://doi.org/10.1002/cssc.202000812

[166] Keyzer E.N., Matthews P.D., Liu Z., Bond A.D., Grey C.P., Wright D.S., Chemical Communications, 53[33] 2017, 4573-4576. https://doi.org/10.1039/C7CC01938F

[167] Lipson A.L., Kim S., Pan B., Liao C., Fister T.T., Ingram B.J., Journal of Power Sources, 369, 2017, 133-137. https://doi.org/10.1016/j.jpowsour.2017.09.081

www.ingramcontent.com/pod-product-compliance
Lightning Source LLC
Chambersburg PA
CBHW071717210326
41597CB00017B/2515